Hello A

This is my second book and another chance to reach those of you who are about to make some unwise choices and costly mistakes in building or remodeling your home. So as not to repeat myself, get a copy of my first book **Home Improvement (Homeowners Most Often Asked Questions)** and read it, especially chapter 1 that deals with choosing a repairman. The same holds true when choosing a contractor but maybe a hundred times more important.

In this book, I give you a step-by-step explanation of why and why not to do something as well as personal interviews with others so you might identify with their successes and failures. We all know that it costs a lot money to build or remodel and if you can do it yourself and save money, I say do it. Just read this book and then decide, please.

My radio program in Houston, Home Improvement Hotline on KTRH, has been on almost 9 years and I am now a regular on Channel 11 TV as well as host of a 1-hour weekly show over HGTV that is shown in over 50 cities around the country. I'm told that this sudden boost to my career is because I've survived these past 9 years. I attribute it to the fact that I enjoy what I do, that I stay current with new information on building and supplies, and I tell you the facts. Truth is always victorious.

Tom Tynan

BUILDING & REMODELING

with

TOM TYNAN

A HOMEOWNER'S GUIDE TO GETTING STARTED

SWAN PUBLISHING
TEXAS • CALIFORNIA • NEW YORK

Author: Tom Tynan
Publisher: Pete Billac
Editor: George Tynan
Layout design: Sharon Davis
Graphic consultant: Paul Titterington, architect
Cover photo: Jennifer Lamb
Cover design: Mark Fornataro

Books by Tom Tynan:

VOL 1 HOME IMPROVEMENT, *Homeowner's Most Often Asked Questions*

VOL 2 BUILDING AND REMODELING, *A Homeowner's Guide to Getting Started*

VOL 3 BUYING AND SELLING A HOME, *A Homeowner's Guide to Survival*

VOL 4 STEP BY STEP, *15 Energy Saving Projects*

Copyright @ April 1996
Tom Tynan and Swan Publishing
Library of Congress Catalog Number 94-65636
ISBN 0-943629-14-4

BUILDING AND REMODELING is available in quantity discounts through: SWAN PUBLISHING COMPANY, 126 Live Oak, Suite 100, Alvin, TX 77511. (713) 388-2547 or FAX (713) 585-3638.

Printed in the United States of America

DEDICATION

To my sister, Theresa. Thank you for giving me new direction. You showed me what my goals, responsibilities and priorities should be during a time in my life when I was lost. Life was too easy for me and you showed me how quickly it can change. May God bless and protect you. I love you, sis.

INTRODUCTION

I don't imagine any of you "local people" reading this book are unaware who Tom Tynan is. You've heard him over KTRH 740 AM in Houston each Saturday and Sunday. He is now on three days; Friday from 10am until noon and weekends from noon until 3pm.

He also has a regular show on Channel 11 television here in Houston from 9 to 9:30 on a Saturday morning plus several other times during the week. And, his OUR HOUSE series is shown in several major cities around the country as well as his HGTV (Home and Garden Cable television) that now plays in over 50 major cities in the United States. Or, you were taught by him at Houston Community College from 1986 to 1991.

Because of his television appearances and his picture on the books, Tom has become highly recognizable. People always stop, shake his hand and compliment him. They ask him questions you wouldn't believe and he answers them with ease, candor and honesty. I am constantly in awe over the fact that he knows so darn much. As Tom's publisher, golf partner, and friend, I've come to know more about him during these past few years.

Tom has a "way" about him, a *charisma*, that makes you like him instantly. He is unpretentious, a *regular* guy with a friendly smile, a memorable, clear speaking voice, and a vast knowledge of the building business. His advice will save all of you *some* money, and many of you a lot of money.

His first book, HOME IMPROVEMENT, has sold over 200,000 copies in two years—mostly local sales. I have over 200 letters from those who read his books telling me how the books have saved them money. That first book saved me over $4000 when a repairman insisted that I needed a new air conditioner. A drip pan for $47 fixed the problem.

I recall how excited Tom was with his first book; all first-time authors are. I called him the day the books arrived at our warehouse and he came out to help unload them! He's a former athlete and works out regularly four times a week. He doesn't mind work.

He unloaded most of the 5000 original copies in boxes of 84 books that weighed about 50 pounds each. He carried three boxes at a time! When the freight truck pulled away, we then sat down and opened a box. I watched his eyes as he looked at the cover of his book. It made all of my efforts in publishing it worthwhile.

He drove 60 miles from his home to help unload the second printing of 10,000 copies! He was in California when the third printing was delivered but I know he'd have been here lending a hand. The ninth printing is due in about a week. Let me make a note to call him.

Tom has been in the building business for just over 15 years. Recently, he got involved in building again but only on an advisory basis. He has his degree in architecture from the University of Miami.

Tom's third book, BUYING AND SELLING A HOME, has had phenomenal sales, many by experienced Realtors.

It tells everything you need to know to **sell** your home in the quickest amount of time with the least expense. And it tells you how to **buy** a home wisely and how to make certain it is as it's represented.

Working with Tom is a pleasure. He laughs a lot (mostly at his own jokes) especially over the air - you've heard him. He doesn't type very well so he tapes his book and I do the typing. We sit down together for hours at a time with him looking over my shoulder correcting things I copied incorrectly.

His popularity is so fast-rising that he's had to get an unlisted number to limit the calls because Tom, in his own sweet, giving of whatever free time he has, answers each and every call with extreme patience and concern. He is an outstanding individual who knows what he's talking about and loves to share it with his viewers and listeners.

In this book, BUILDING AND REMODELING, Tom tells everything you need to know to decide whether to build on your own or hire a contractor. He tells about small remodeling jobs and how to go about doing them and he shows you actual sets of house plans and specifications, contracts to sign, and insurance papers to look for.

When he endorses a product or service, you can be certain that the company and the product is determined to be good. Not for any price will he go against these convictions. If he mentions the name of a company or lending institution or a person, it's because he has, personally, had good business dealings with them. He doesn't guarantee that your relationship will be the same, only that his was.

Tom and I have spent weeks on the rewrite of this book and (with Tom's dad, George, as editor) it's finished. Tom is now making notes on a fifth book, either one where parents and children can build fun things together or perhaps another book of questions and answers. We plan to have it out before Thanksgiving.

Tom's fourth book, STEP BY STEP (15 Energy Saving Projects) is in print now and is being scooped up by energy companies around the world! They like what he has to say so much that I've had to print copies especially for them. These companies are buying thousands of copies at a time to give to their customers. Yes, Tom Tynan's popularity is growing rapidly.

Enough of an introduction, let's get into this book and see how much you learn from it.

Pete Billac
Publisher

TABLE OF CONTENTS

Chapter 1
DO YOU HAVE THE GUTS 15

Act as Your Own Contractor
Don't Lose Your Cool
Is It Worth It
Homebuilder #1
You Can Do It
The Birthday Surprise

Chapter 2
BUILD OR REMODEL 30

Set Your Goal(s)
All About Financing
Checklist for Remodeling Loan
New Home Construction Loan

Chapter 3
ROLES PEOPLE PLAY 42
Architect/Designer
Engineer
General Contractor
Banker
Building Department and Inspectors
Civic Associations
Friends
You

Chapter 4
PLANS AND SPECIFICATIONS 53

Site/Roof Plan
Floor Plan
Exterior Elevation
Foundation Plan
Wall Section
Interior Elevation
Other Details

Chapter 5
SETTING A BUDGET 67

Pad and Pencil
Use the Telephone
Remodel Completely or Partially
Adding a Swimming Pool
Cut Sheet Information
Decisions—Decisions—Decisions
You Can't Do It All Today
It's a Mess
Sample Cut Sheet

Chapter 6
PERMITS, INSPECTIONS AND CODES 80

Pulling Permits
Underground Plumbing Inspection
Foundation Inspection
Framing Inspection
HVAC (air conditioning) Rough-In
Plumbing Top-Out

Electrical Rough-In
Final Inspection
Handling Inspectors

Chapter 7
BEING, OR HIRING A CONTRACTOR 90

Are You Legal
Do You Have the Time
How Much Will You Save
Do You Have Organizational Skills
You Can Make it Happen
Can You Handle the Pressure

Chapter 8
HOW TO SOLICIT BIDS 100

Finding Subs
Getting Bids
Things to Look for in a Proposal
Comparing Bids
Choosing Subs
A Golden Rule

Chapter 9
SEQUENCE OF EVENTS 117

The Pre-Planning Stage
Plan of Action
Final Checklist
Remodeling a Bath

Chapter 10
MOST COMMON MISTAKES 125

Not Pulling Permits
Insufficient Plans and Specs
No Detailed Cost Estimate
Going With Low Bids
Using Substandard Material
Project Too Big For Your Skills
Making Too Many Changes
Underestimating Cash Resources
Listening to Bad Advice
Tom's Secret Formula

Chapter 1

DO YOU HAVE THE GUTS

ACT AS YOUR OWN CONTRACTOR

This chapter is going to tell you some of what you might expect if you are going to build on your own or act as your own contractor. The first thing I want to tell you is, *Don't be intimidated by building on your own! Most people don't feel they have the skills to pull it off, but they do . You do!*

Everyone practices these skills every single day of their lives. If you can run a household, office, classroom, business or just your own life, you can run a job. The whole key is to learn how things are done, then organize them. I'll tell you how they're done in this book.

What I recommend you do if you are planning a project of any size, is to stop sitting around watching just *anything* on TV and try to find a home improvement show to watch (mine). Or, take a class on being your own contractor. Read books about home repair, (mine). Listen to home improvement shows on the radio (mine).

If you're an average person without any building skills, I promise you can do it. Educate yourself first then start. Before long you'll feel the confidence creep, then rush into your body and it will become easy for you. That's how I started.

DON'T LOSE YOUR COOL

To begin, I'm going to assume that most of you are married and that you'll be building this together. I hope you have a solid relationship because there will be times when things don't go exactly as planned and people tend to lose their cool. Even if you're single, dating, or engaged to be married, there will be uncomfortable parts of this building business, seemingly insignificant things that add up.

Those of you who listen to my program know that I have a weird sense of humor. I'm told I make jokes and laugh at them myself and that the jokes aren't even funny. The fact is, approach this building-yourself-project with a positive attitude and a sense of humor. Here's why.

You'll eat a lot of cold sandwiches, pizza that becomes stiff in the sun and "cold" drinks that are hot. You'll need to take frequent baths and showers because you'll get dirt and sawdust in your hair, ears, nose and every other place on your body.

Your patience, tolerance and temper will be tested frequently. Handle problems as they come. If your sub-contractor doesn't get his job finished and you've already called the building inspector who is driving up, you can't kill the sub. You can't even hit him with a board with a large nail in the end. You just talk to him and try to reschedule the inspection.

For some unknown reason, you always run out of time in getting something finished. You've been working as fast as you can from dawn to dusk but you don't have the 15 or 30 pound felt on the roof. There's no worry, the weatherman said it wouldn't rain. But, it **does rain!**

The electrical tools develop some rust, so clean and oil them. Some inside lumber warps, but only a few boards, replace them. The bags of cement you forgot to cover is now three tiers of stone, so put a table top on it. And the only dry spot is where the hardened pizza kept the rain from reaching the wall outlets in an opened box. Don't eat the pizza.

Don't plan on encountering many honor students in the general work force, either. I'm not saying they are idiots, just not *summa cum laude* graduates of Harvard. There will be words and phrases that are new to you. As a first-time contractor, take it all in and smile, it's the "building people's" vocabulary. There's a well-known, self-explanatory term most people know called "plumbers crack". Whether you look for it or not, it's there on most overweight workmen.

I have one sub who is too good to let go but he starts off each day with his pants around his waist, and at the end of the day they're near his knees. I'm "mooned" from about 10 am on and it just isn't a pleasant sight. Sorry, if I've offended any of you; these things are real! I see it all the time. You'll see it, and now that I've made you aware of it, you'll look too.

Married or single, your love life will suffer! Husband and wife working together is not conducive to romance, not on a building site anyway. You ask her to get something and she doesn't respond. You ask a second time and she looks at you like you're an **alien.** The third time you shout at her. Maybe you forget about it the next day but she will **never** forget!

If you want her help ever again, better start practicing your apologies, explain or show her what you want or get it yourself. But, don't ever do it again! She might not know all the fancy new tools but there are things she does know that are foreign to you too. Yeah, keep cool.

Keep pets at home, away from the building site. One guy nailed his wife's cat in a bathroom for a holiday weekend. The cat lived but wasn't that happy, and for months, until the home was finished, visitors knew where the bathroom was without looking, depending on which way the wind was blowing.

Neither of you will enjoy breaking out your checkbook to buy things that were lost, stolen, broken or that you forgot to include in your plans. But it happens on every job. Plan for some errors, lost or forgotten items when doing your budget. Don't be surprised when they happen, plan on them happening!

Forget a social life. You might take time off and go someplace and your body will be there but your mind will be back at the building site. You won't get much sleep either and you'll forever feel tired. Getting up a few hours before your normal workday, eating a Big Mac for lunch while driving to look again, and dropping in a few hours after work takes it's toll on the body.

When walking through or working around a building site, wear a hard hat for when the carpenter or roofer drops their hammer. Wear cotton gloves to guard against splinters because when you least expect it, you'll pick up a seemingly smooth board and a splinter gets under your fingernail and it hurts!

Watch where you walk. There might be wood with nails sticking up, you could trip over a multitude of things or fall in a hole dug for plumbing or worse, a deep hole for a septic tank.

And leave your Dockers at home; this is no fashion show. Either drag out your worst, old clothing that is loose fitting and comfortable or spring for some coveralls. If you have only new, comfortable clothing, it gets old quickly. Get a hat too, hair or not, that sun will bake you.

IS IT WORTH IT?

There are any number of situations that come up that you'll learn to handle, all unpleasant ones. You'll look in the mirror and just shake your head for ever starting such a project. No matter how cool you plan to be, you'll lose it a few times and feel bad about it later. Somewhere along the way you'll wish you never even heard the word "building" or "contractor" or "home".

But, is it worth the trouble? When it's finished, it will be!

Your answer will be an unequivocal YES! Ask anyone who has ever built or remodeled a home. Most will say they'd never do it again but the feeling of accomplishment is unparalleled.

Let's learn how to do it right. Let's plan and budget and follow a program. Actually, when you finish this book, if you follow what I say to the letter, you could do your own building. I'd prefer that you read and learn more, and attend a few classes given at most local colleges, but if not, wing it and see how much fun it can be.

Make a game out of it! And, enjoy the game! Confront those expected problems head on, and treat the unexpected ones as if they were expected! You can do it, I promise you, you can! But to stack the cards in your favor, read my book and follow what I say. Visit building sites. Help a friend build their home, Learn - then practice, practice, practice.

HOMEBUILDER #1

This is an interview I conducted with David Russell, vice president of Texas Commerce Bank. David's house is one of the most successful homes I've built, and is so energy-efficient that the local utility company is monitoring the electric usage to find out why such a large house doesn't use more electricity.

David has a personality somewhat like my own. He compares building to childbirth. "It begins with some discomfort, pain and suffering, sleepless nights, anxious moments, last minutes rushes, perhaps scars and stretch marks, yet many women want to repeat the procedure a year or two or so down the line."

David, being a man, hasn't actually given birth to a child but he's been what you might call the helper, the sub-contractor, three times. His wife was the builder. Perhaps we should interview her, but she is unavailable, busy watching over and running behind the kids.

Q.   David, would you build another house?

A.   Yes! It was a challenge but it was fun and I felt good when it was over. On my next home, I think I could do it faster and easier, maybe even save a few bucks. There are also some changes I'd make.

Q.   What would you do differently?

A.   Nothing profound. I wouldn't change the size of the house (it's a monster at 4000 square feet) but I would add on a room or two and make a few of the others smaller or larger. For instance, with kids, the utility room needs to be three times the size it is.

It happens this way with everyone. No matter how often they go over the plans, they want changes when it's finished. Hindsight is always 20/20.

Q.   You mentioned having nightmares, David. Want to tell us a few?

A.   Well, my builder (he's talking about me) always told me that anything can be fixed and, unfortunately, I never took that to heart. One evening I came home to look over the job and discovered that there was no light switch at the bottom of the stairs.

This was critical! I didn't want to say good night to the kids and then have to turn the switch out before I went down the stairs or leave the light on all night. And my wife would have to do the same thing, sometimes carrying the baby in her arms.

I laid awake all night long thinking of this problem, certain they'd have to tear out the sheetrock and wallpaper and patch it and it would look horrible, or at best, delay completion of the house a day or so and we didn't have that day. We had to move in as scheduled.

But, Tom assured me it was a minor thing. It wasn't in the plans but Tom anticipated I'd want or need one and the wiring set-up was right behind the wall. They punched a small hole, put in a switch, closed it up perfectly and the problem was remedied.

Q.   That was one. Surely there must have been others?

A.   Another time I lost sleep was when it rained one day after the weatherman said it wouldn't, and lots of things got wet on the inside. I was certain the rain would rot the wood and in a year or two the house would fall apart. But the rain did minimal damage.

And there was a problem with the fireplace; the room was rather large and we wanted a certain size firebox. We had given the builder *inside* dimensions, and the subs thought they were from the *outside*, a difference of two feet. I groaned so loudly that the entire neighborhood heard me. I laid awake tossing and turning the entire night on this too but, it was easily fixed.

> Those of you who get involved in building will under-stand that if you plan correctly, anything that goes wrong can be handled. Pros with 30 and 40 years experience have the same problems.

Q.   David, tell us how this all began with you. And, do you feel that just about anybody can act as their own contractor and/or build on their own?

A.   You begin with an idea or a dream, buy some property, have a set of plans drawn and go from there. It's scary at first but the more you get into it, the easier it is. It's your house, you are building for the first time and you want things **exactly** as you envision them. No problem! That's why it's important to be "on the job" to watch over things so they can be corrected with very little effort and expense.

Q.   Would you recommend this to someone with absolutely no experience!

A.   I don't think anyone had less experience with building, but my wife and I wanted a house that we designed. I was fortunate to run into you, who acted as my contractor, but I had to do a lot of the legwork. I put everything down on paper and begin numbering the priority in which things had to be done. It's easy; I think anyone can do it.

Since I built my house (contracted to have it built) I have learned much. If, when we build our new house, I'll know more but I'd change some things I did the first time. I've learned not to panic, or worry so much, because if you take it a step at a time and if you plan your work then work your plan, everything falls into place. Then *voila!* you are living in your new home. It's a great feeling.

Q.   How about your family? Did you feel any kind of strained relationship with your wife while the building was going on?

A.   Tom, there were absolutely no problems of this nature. My wife and I went over the plans, she told me what she preferred and I complied. I then added what I wanted and that, perhaps, is why the house ended up being a bit over 4000 square feet. But she was wonderful! "This is your project baby," she told me. "You know what I want, you know what I'd like, now run with it!"

Her big concern were the trees. We built on a heavily wooded lot and she wanted to save as many of the trees as we could. I think we built around them all! Much of what I thought were headaches or nightmares were from panic. It is really easy and anyone can do it if they try.

Q.   But you made changes in the plans as you went along, didn't you?

A.   Doesn't everybody? Naturally, even after all the planning, looking, discussing, changing and changing again, we'd have done a few things differently. When I build my next house, I'll put in even larger closets that we have because we're both *clothes hogs* and would each like almost a full room as a closet.

I cannot imagine living in the homes of 25 or so years ago that had such small closets. Remember those, maybe 3 feet long behind a door or the master bedroom with one wall of closet behind folding doors? How did anyone live that way?

Would I build again? Of course, I'm going to build again. I can't wait to begin!

THE BIRTHDAY SURPRISE

I interviewed a person who prefers that I not mention his name because he's been secretly remodeling a home that he bought for $62,000 about 7 months ago and he wants to finish it and surprise his wife with it for her birthday. He has 6 weeks to go and agreed to this interview but was *antsy,* anxious to get back to work.

He had never done this kind of work before but he was one of the first to buy my Home Improvement book and it inspired him. He said if the repairs were so easy to do that the remodeling would be likewise. Here are a few questions I asked him. I hope they help you.

Q.   So, my book helped you make the decision to remodel. Was there a lot of work to be done?

A.   Yes there was. I had to put on a new roof, install central air and heat, tear out and remodel the bathroom and kitchen, and make a two-and-a-half-car garage out of a one-car garage. It was a three bedroom home but I cut out one bedroom and a hall and made a large master bedroom, enlarged the bath area and put in two large walk-in closets. What do you think of it?"

Naturally, the first thing I looked for was the continuous ridge vents and, they were there! There were only studs and open ceiling showing because the sheetrock was coming later that day. He and two helpers were laying out the insulation to be installed when I was there. Installing the insulation before the sheetrock is a lot easier than bending over in the hot attic.

The wall wiring and outlet boxes were in, the bath area was being tiled by a sub, someone was working on the kitchen cabinets and everything seemed to be running smoothly. I was impressed, amazed at how well-organized things were.

Q.   And, you've never done this before?

A.   No. It actually took longer to plan than it did to do, Tom. I drew a rough sketch of what I wanted to have done then took it to a student at the college who charged me $300 and did it in a week. Then I looked for subs by calling newspaper ads and that took some time. I got bids, figured my budget on what I could and couldn't do, made a bank loan, hired and fired a few people and here I am.

I have a sheetrock man who will tape and float, the painter is coming over day after tomorrow, the paper hanger two days after that and then I install the carpet and linoleum in the kitchen. I always add one day in after the work is done before I schedule the next person, you know, in case something needs to be redone.

I'd hire this guy in a minute to run any job for me. I had never seen such organization in a first-time builder. He was really surprised to see me and I got a feeling of self-satisfaction knowing that my little ten-buck book prompted him to undertake such a major remodeling challenge.

But friends, when you think about it, this is no different than repairing or doing over one thing at a time when it goes wrong with your house. If you need to add a new bath or make the garage a large playroom or even add a room, don't be afraid. I say do it and accept what comes.

I really love what I do because of people like this one and because of so many of you whom I know are doing these things yourself; saving money, fixing up your home the way you want it and feeling good about what you've done.

This book will help those of you who are "on the fence", wondering if you should take the chance to do this on your own. You know that I "tell it like it is" and there's no bonus for me if you try this and fail. You won't fail! Not if you read what

I have to say and do it in sequence. You don't even have to be smart; look at me, I'm no honor student but I have learned to organize.

Did you get those two key words, "learn" and "organize"? I tell you exactly how to do both in this book. I teach you enough words so that you can ask intelligent questions. The people who are going to sub for you will help and guide you. And organize, that's something you'll have to do on your own but I tell you how to do that too, one step at a time.

It doesn't take guts to do your own building or contracting. It just takes sitting down and putting a pencil and paper to it, to list what you want done and then to get bids to see if, financially, you can have it done.

I wouldn't lead you wrong. Those of you who hear and watch me on these shows know that I take pride in giving advice Time and time again, I hear stories like the one above on someone who went out and tried it. And the guy above did it while reading my Home Improvement book where I answered questions homeowners ask.

This book tells everything you need to know to begin building or remodeling on your own. If you're afraid to try, look what the guy above did. He did it like I say to in this book. But by reading my first book, he decided he could do it. If he did it with that basic book that mostly told of minor problems, he could build a city after reading this one. And. . . so could you!

But, no need to decide now. Read each chapter so you will be able to determine whether you want to nor not. There are problems to contend with such as time, financing, budget, how quickly you need to have it done and a few others. To tell you the truth, I'm really proud of this book. It makes me feel a bit more cerebral than answering simple questions. I like them both but this one, I feel, gets down to those who are adventuresome, those of you who have the guts to try?

Chapter 2

BUILD OR REMODEL

The main question most people have to answer is whether they should take their old home and remodel it, or just fix it up and sell it and build a new home? If this is decided and they opt for selling, should they buy an older home and remodel it in lieu of building a new home?

Regardless of what you'd like to do, drag out that checkbook and see what it is you can do. Let's assume you have no money problems. So now, it's your personal decision to build a new home or buy and older one (many people are doing that) and remodel it.

My suggestion is that you deal with this problem academically. Get a pencil and paper and see what the cost will be to remodel the home you're now living in and whether the cost of the work justifies itself, you know, whether the value of the home is enhanced at least by the amount of money you're putting into the project.

Go to a local real estate office to get comparable prices on other homes in the neighborhood; this should help you decide. Ask them for a printout on everything that's been sold during the past six months in the neighborhood so you'll have an idea what your home is worth. They'll do it! If not, tell them you're planning on selling your house and that they will get the listing.

Other things you need to consider is the emotional and convenience aspect. Do you like the neighborhood? Do you like your neighbors? Do you like the schools? Are you close to where you work? How about shopping, distance to the freeway, is the neighborhood safe? Do you want to pack those few hundred boxes to begin moving?

While you're planning on remodeling your present home, you have a deal on a large lot in a nice neighborhood only a few miles away. And you saw and old house for sale in the other direction that looks like a winner; perhaps you can remodel it. Decisions, decisions, decisions. What to do? Sit down and think it out.

SET YOUR GOAL(S)

On remodeling your old home or a new "old" one, determine what your goal is and what you would prefer your home to be like. Do you need or want a larger bedroom, an extra bedroom, larger utility room, a three-car garage, things like that. Get the entire family involved for their opinion and vote in the matter and give their input consideration. You're a family, right? Then, get the figures on what it will cost.

Some of you will think this is basic common sense and I don't mean to insult any of you but it just so happens that many, from my experience, although they might know about this, few do it. The biggest mistake most make is feeling that the size of their home is not sufficient. This might be fact, they might need a larger home.

You might have a 2400 square foot home and want one that is 3500 square feet but friends, with most homes, it isn't the *size* that's the problem, it's the utilization. Things are just in the wrong place in the home you're living in now. Your wife would prefer that the utility room be larger, the 2 boys want their own bedrooms, you'd like a larger den with a wood burning fireplace, your daughter wants her room to be larger and would prefer it face the street, and the dog wants an entry/exit door so he won't have to sleep in the un-air conditioned garage on hot summer nights.

So, look at each aspect of your present home to see if you can do without the dining room; you all eat in the den anyway while watching TV and the breakfast room is kinda' fancy. Nobody uses the living room except the dog when it's asleep on the couch so you can do without it; let the mutt sleep on a blanket. And mom's sewing room is being overtaken by dust and cobwebs because she hasn't sewn a button since dad got his promotion 9 years ago. You can take that room and put one of the boys in it!

Maybe you don't need a large home. You like this neighborhood and you can probably remodel. Why not make the changes and come out far cheaper with the home you're living in now? You're not *that* concerned over a little money but you are over a *lot* of money. Let's see what it costs to change this place as compared to purchasing a new one. Where do you begin?

ALL ABOUT FINANCING

I interviewed my banker, Steve Rife, vice president and loan officer of Sterling Bank. I asked the questions you need answers to and thought it best to get them from the "horses mouth" instead of from jackasses who don't really know what they're talking about.

Q.   Will the bank lend money to a homeowner on the value of his home?

A.   Yes. Banks in Texas are governed by the state Homestead Laws and are allowed to make home improvement loans to a homeowner on the value of their home for three things; to build a house, to remodel a house, or to pay taxes on a house.

Q.   How about remodeling loans? Tell me about those.

A.   We love remodeling loans because we are dealing with a homeowner, usually a person with a family, steady job, responsibilities, you know, someone who is stable.

If a person wants to build a room that's 3 feet wide and 10 feet long, from the banker's viewpoint, this won't add value to anything. It might make it more convenient to you but to banks, we look at how it will increase the value of your home so we can justify lending money against it.

If you're going to put a fishpond in your back yard, I don't get very excited over that either. But if it's a pool, we take another look at it. As your banker, and since the bank loans against value, I need to make certain it enhances the value of your home and it depends on several things; your neighborhood for one.

If you have the most expensive house in your neighborhood and you want to add a swimming pool that will cost $25,000, I will go in and ask if you feel it would be wise to build a new house in another neighborhood that includes a pool. I don't mean to act as your critic but it's my job to tell my customers about that. Some don't appreciate the intrusion at first but I explain.

Because, a pool isn't all they'd like. They also want a bathhouse, an underground watering system, a *porte-cochere* on the side of their house and it goes on and on. So I say, in the best, not know-it-all manner I can muster, "Folks, this, in my opinion, puts you so far above the value the area warrants. It is not a good loan for the bank and not a wise investment for you. I suggest you weigh this out and decide."

You see Tom, (and those who plan to remodel in this fashion) from the bank's viewpoint, we look at it from the worst case scenario. Suppose we had to take the house back for any number of reasons; unemployment, bankruptcy, divorce, relocation, death by the main income-earner or any other catastrophe. Our goal—our business—is to get our money back, hopefully on time and with interest.

On home improvement loans we loan 100%. We look at these loans closely because as secondary lien holders, if the improvement doesn't enhance the value so as we're covered, we pass it up.

CHECKLIST FOR REMODELING LOAN

1. Copy of your Deed of Trust
2. Copy of your Title Policy
3. Homeowner's insurance papers
4. Flood insurance (If applicable)
5. Bid, proposal or contract for the work to be done.
6. Three (3) year tax returns (if self-employed)
7. Application, signed and dated, by all parties on the Deed of Trust
8. Proof of paid property taxes/Copy of assessment for previous year.
9. Name and address of first mortgage holder and year end balance of first mortgage loan.

These are the items you will need. Do yourself and your banker a favor and follow this list to the letter. If you go in with this information in a neat folder and hand it to your banker, I promise you they will be impressed. When you get the loan and everybody's happy, tell them you got this information from reading my book.

NEW HOME CONSTRUCTION LOAN

Now, how about a loan for new construction? If you decide to build a home on your own, and the first thing I recommend is that you talk to your banker.

Remember this, please. Your bank is in business to make money. They like to make loans on homes that are sensible and that the homeowner can pay for. They will loan 80% of the appraisal or of the contract price, whichever is lower.

Here's a list of items you need to impress your banker with when you apply for a new construction loan.

What you need at first:

In your initial conversation with your banker to inquire about a loan, I think it would be wise to have a copy of your preliminary plans (that consist of floor plans and elevations) for them to look at. They are far less expensive than a full set of plans and you need these anyway. If it seems reasonable to the banker, they will give you an application to take home and fill out.

For your first real visit, here is what you need:

1. Completed Loan Application.
2. A copy of a full set of plans and specifications.
3. Copy of Contract from your builder.

4.   Earnest Money Contract. If your permanent loan is with another lender other than the bank that is making your construction loan. Nor do all mortgage companies provide construction loans, Some banks, however, handle PERMANENT and CONSTRUCTION loans in one package. Yet another reason to talk with your favorite bank and banker first!

The bank usually has its own appraiser. They give this appraiser the information you provided (plans and specs) on which to make its decision.

On your final trek to your bank, you'll need:

1.   Your "take out" Commitment Letter from your Mortgage Company.
2.   A survey of the property.
3.   Title Binder (that is provided by the title company.)
4.   Builders Risk Insurance (provided by the builder).

Q.   What if you buy an existing house and want to take out a home improvement loan to remodel it?

A.   The norm is that you need to be in a home at least 2 years before banks make a loan on it. Chances of a home improvement loan are not good. Since the original loan was at the limit, we might end up loaning 90% of value; not a good business decision.

IMPORTANT

THIS IS SO VERY IMPORTANT TO THOSE OF YOU WHO ARE BUYING AN OLDER HOME AND PLANNING TO REMODEL IT OR JUST FIX IT UP TO MAKE IT LIVABLE. I I FEEL THAT IT IS SO IMPORTANT THAT I'M MAKING IT IN VERY LARGE LETTERS TO MAKE CERTAIN YOU READ IT. PLEASE READ IT CAREFULLY—IT'S SIMPLE BUT VERY FEW PEOPLE ARE AWARE OF THIS!!!!!

I SUGGEST YOU GO TO THE BANK BEFORE YOU BUY AN OLDER HOME THAT YOU WANT TO FIX UP. THE MORTGAGE COMPANY (OR WHOEVER THAT LOANS THE PERMANENT FINANCING) HAS LOANED TO THEIR LIMIT SO IF YOU COME TO US, YOUR BANK FIRST, WE CAN TALK IT OVER AND SEE IF IT WILL WORK.

FOR COUPLES WHO ARE PLANING TO BUY AN OLDER HOME TO REMODEL, LET'S SAY THEY ARE BUYING A HOME IN AN OLD NEIGHBORHOOD FOR $50,000 AND THEY WILL NEED $30,000 TO FIX IT UP WITH CENTRAL AIR, STORM WINDOWS, A NEW ROOF, ETC., I CAN WORK IT THIS WAY TO THEIR ADVANTAGE.

I WILL LOAN THEM THE $50,000 TO BUY THE HOUSE AND ALSO THE $30,000 TO FIX IT UP ON A SHORT-TERM LOAN. THEN, THEY CAN GO TO A

MORTGAGE COMPANY FOR A PERMANENT LOAN. WHAT THIS DOES IS, INSTEAD OF GETTING A 30 YEAR MORTGAGE FOR $50,000 AND A HOME IMPROVEMENT LOAN FROM THEIR BANK, (THAT IS PAYABLE IN 12 YEARS AND HAVING TO WAIT 2 YEARS TO DO IT) FOR $30,000, THEY HAVE THE LOANS CONSOLIDATED.

THIS ENABLES THEM TO MAKE ALL THESE IMPROVEMENTS NOW AND NOT HAVE TO WAIT TWO YEARS. THEY CAN THEN TAKE THIS BANK LOAN TO THE MORTGAGE COMPANY FOR A 30 YEAR PERMANENT LOAN. IT'S SMART BECAUSE IT LOWERS THEIR PAYMENTS AND MAKES IT EASIER TO RESELL.

THE BANK IS IMPORTANT TO THEM AND IF THEY WILL ONLY ASK AND DISCUSS IT WITH THEIR LOAN OFFICER AT THEIR BANK, IT MAKES A DIFFERENCE. WE ALSO HELP THEM WITH THE MORTGAGE COMPANY AS MUCH AS WE CAN AND WE GUIDE THEM CORRECTLY.

The above information is the SMART way to do this. Many people feel banks are for checking and savings accounts only. A good bank and a good banker will save you lots of money, as per the example above. I've done it this way many times and it was smooth and easy.

Q.   What are the criteria for a homeowner to qualify for a home improvement loan?

A.   We get as much information as we can on their financial position. Their salary, savings, debts such as loans, credit cards and charge accounts. If that passes, we order a credit check on them.

We then look at their stability in their job or profession such as how long employed with their present company. If someone changes jobs frequently, we ask why. If the answer doesn't suit your banker, chances are you won't get the loan. I might ask for tax returns or get permission to talk to your employer and former employer.

If a person wants to get a small home improvement loan fast, like maybe the air conditioner breaks down during the hot summer months (when else do they break, huh?) and they want to bypass the time it takes to qualify, I will take any collateral (like a car) and instead of charging for a home improvement loan, I will give them a form which enables them to write the interest off on this loan as a home improvement loan and get them the money right away.

I f you have **bad credit**, if you **owe** more than you **own**, if you have a job only a short time, or haven't established credit, I recommend you wait until you work things out! I don't want you to get your hopes up or waste any time or money trying to build a new home and get a loan from any lending institution. Your chances are about as good as a snowball in . . . Houston in July!

Regardless of your present situation or how bad you are financially or jobwise, it's worth an hour or so of your time to visit your banker. Laws and rules are always changing and you can meet your banker and find out the criteria for making a loan.

This interview I had with Steve Rife from Sterling bank should help you. I just wish each and every one of you had a banker like Steve. He's my banker and worth the trouble to drive from wherever you are to do business with him.

Chapter 3

ROLES PEOPLE PLAY

If you are building a house that costs $50,000 or $1,000,000 the procedure is exactly the same. Many people are involved and here is a list of the major players.

ARCHITECT/DESIGNER

The first *player* many tend to ignore is perhaps the most important; the architect/designer. A recent survey by an architectural magazine showed that homeowners feel these people are needed the least. They, however, are one of the most important!

They produce the plans that serve as the contract documents that protect you during your **entire building process**. This is why your architect/designer should be skilled in all facets of building.

Architects/designers are paid in many different ways. The AIA (American Institute of Architects) say an architect should be paid 4% to 10% of the cost of whatever is being built. Most that I've found charge by the square footage of their drawings. For a 3000 square foot home they charge $2 a foot or $6000 for the plans.

I recommend a flat fee. You should talk this over with your architect. Doing a percentage like the AIA wants is, I feel, fine for large buildings because they have a full staff of qualified people necessary to build a skyscraper, but a house is no great feat such as this. I like the flat fee also because the architect will work within your budget.

The difference between an architect and a designer is that the architect has a college education and has passed the training requirements as well as a written exam registering him in the state in which he practices. Many designers are also college educated, have passed tests and requirements, but work primarily on residences. Few work commercial buildings. Some states require that a registered architect put his seal on all the plans.

My choice is to find someone who knows what they're doing whether they've been to school or not. I'd prefer an architect or designer who has actually built what I'm planning to build; not someone who sits and draws lines and puts numbers down. I've run across dimwit architects and unqualified designers and thankfully, also many skilled ones.

In determining the qualifications of either, I'd ask what each has done and look at several of their finished products. If you see a particular recently built home that you like, knock on the door and ask the homeowner who designed it for them. It's not a bother to them. I've found the homeowner feels complimented and will gladly give you the information.

Sometimes, designers and architects have plans to a home they built for someone else that you like and they can offer these plans to you at a discount. You see, an architect owns the plans to your home and they can sell them to whomever they choose. Hopefully, it's not your next door neighbor.

There are companies who sell house plans that you can purchase at a reduced rate but know that it takes time and skill and a substantial effort to design a home for a certain individual. The designer and/or the architect earn their money.

ENGINEER

The second person you'll deal with is the engineer. The word "engineer" will appear on an architects plans. This doesn't mean an engineer actually drew the plans, the architect does that. But architects need to work with an engineer who comes in on residential building for foundation design. Most architects have engineers they work with on a regular basis.

You'll need an engineer if you're building on a slope or on the side of a mountain or where you build higher, wider or more unusual that what we might call the "standard" homestead. Also, if you have a large span, a beam, if there are soils in parts of the country that need special care, the engineer will specify the type of concrete or steel you'll need.

The architect designs the house first. Then he designs the foundation. He then takes the plans to the engineer who checks the details to approve or correct the requirements. This then goes back to the architect for changes. Oftentimes, the engineers seal is necessary to *pull* your building permits.

GENERAL CONTRACTOR

The third person in the scheme of things is the General Contractor. The architect makes several sets of plans which he hands over to the General Contractor, you know, the head guy who supervises and coordinates it all. Then the general contractor gets these out to his favorite sub-contractor(s) to bid.

Contractors are paid in different ways. Some ask for a percentage (as much 20% of the building cost) and others will give you a flat rate to build the entire project and squeeze as much profit out as possible.

The way I prefer is the way I recommended with the architect; a flat fee plus the cost of the job! This is known in the building business as "cost plus".

After you've chosen your contractor, sit down with him and ask his opinion as to changes, better methods of building, stronger or less expensive material or new and innovative things on the market.

BANKER

In working with a banker, the first thing to discuss is a construction loan; a loan whose funds are used to build the house. Construction loan financing rates are usually higher than permanent financing but they are short-term loans. You use the money only as you borrow it; some for the foundation, some for the framing, the roof, etc.

Most banks, currently, are working on a 80%-of-cost loan which means you come up with 20% and they give you the final 80% and the money is there for you to use whenever it is needed. These are things you need to work out with your banker.

The banker schedules inspections to see that the money you draw is being paid out. They see one part finished, inspect it and you get the money to pay for it. Try to find a person who will be handing out these draws that you feel compatible with because they, in essence, will be your building partner.

In choosing a lending institution (bank, savings and loan, mortgage company) go first to the one that you're familiar with or one that is recommended by someone whose opinion you respect. Go to at least three. Play like you're going before Judge Wapner on *People's Court* and he wants you to have at least two, preferably three estimates.

You need to be satisfied with the rates, the terms of the loan and the person you'll be dealing with. Remember, this person will work closely with you and the general contractor making sure that completed work has been paid for.

BUILDING DEPARTMENT AND INSPECTORS

These are the people who issue the permits for you to begin building. You must complete all of their requirements before you start. Your contractor usually pulls all the permits for the job.

The building inspectors will come out to approve (or disapprove) what has been done. If it doesn't pass inspection, it has to be done again—correctly!

Some home builders don't like to have something done over again because of the delay and maybe a slight boost in the cost, but think of these building inspectors, also, as partners. They help you. They are there to see that what is supposed to be done, is done! Look at it as a positive; you have yet another group of people watching out for your interest.

If you're acting as your own contractor, you pull these permits (I say "pull" which is the vernacular used in construction. It simply means to get a permit in your hands). You can also hire someone to get these permits for you who know the inspectors or the office people and they can usually get it done faster.

You can find such people (and companies) by asking the secretary who answers the phone at the building inspectors office, perhaps through your building association, from your architect or a builder; anyplace builders hang out. If you're at a local bar, ask in there. Somebody will volunteer the information.

I say be thankful for building inspectors. Most have been builders and are trained to look out for you. I've always gotten along well with building inspectors wherever I built. Treat them with respect and they will, in turn, do likewise and help you in every way they can. They are, really, nice human beings.

CIVIC ASSOCIATIONS

Many planned neighborhoods have these groups to

make certain that what you build is complimentary to what is already there. These folks are voted in by other neighbors, oftentimes those who don't want to do the jobs themselves. I've seen good and I've seen bad.

Sometimes, a person who is "in power" really exerts this power without thought and can be a problem. One homeowners group at a condo suggested that a man remove his screen door. When their suggestions and warning weren't heeded, they physically took the door down themselves. The irate condominium owner took them to court, and over $20,000 was spent on legal fees to satisfy a $75 screen door.

You can't blame the lawyers on this one; it was the dumb homeowners committee that wouldn't make an attempt to compromise and a hard-headed tenant who's pride was severely wounded, feeling his rights had been violated.

What I suggest you do with such a committee is bring them the preliminary drawings to whatever you plan to build and get their approval. If they have the "Little Caesar" complex you will flatter them if you ask their opinion.

But, no matter how nice you are, the good and the bad committees have a job to do. If your planned dwelling is not up to what their organization deems fit for the area, you can make the adjustments during the preliminary stage; modest changes and modest charges.

Don't fight these people, unless of course, you and your architect agree that their demands are absurd and your rights have been violated. Most of the time you'll have no problem. They're homeowners just like you and they aren't getting paid for this chore. If they protect their neighborhood from your unsightly or bizarre plans, they'll do the same on others and protect your investment too. When they do approve it, get it in writing from them and put it with your plans.

Every neighborhood has a royal pain-in-the-neck who will object to whatever you are doing. And they have a friend, also a royal pain, who agrees with them. They will watch what you're doing and talk to you about it. Be nice, whip out that agreement by the civic group or neighborhood committee and you're home free. Avoiding problems is so much smarter, and oftentimes even more satisfying, than battling it out with one of these trouble- seeking busybodies.

> **What I think is the perfect deed restriction is, "Build what you want as long as it isn't ugly!"**

FRIENDS

One big problem with building for yourself or even in using a contractor is the advice you get from friends, neighbors or relatives. I say this because I have had to fight this problem far too often on jobs I did for people.

This advice, though well-meaning, is usually both incorrect and a setback. If you speak with two of your own brothers, three friends and two neighbors, you'll get seven different opinions or suggestions.

Listen to your contractor. Friends and family might love you but if they are not in the building business, they will only confuse you. If they are in the business, listen. The more you know the wiser decision you can make.

Can you be the architect/designer? In most cases you cannot, so don't even try! You can be the engineer only if you *are* an engineer but if you're an electrical engineer, let the foundation engineer do his work and suggest things only in your realm of expertise.

Ask your architect or designer questions and listen to their answers. They will guide you correctly. If you want a skylight here or a high ceiling in the den, they will tell you why or why not it will work. They will design and build whatever you tell them to if it's safe and within your budget.

How about being your own banker? We'd certainly all like that one, wouldn't we? I served as my own bank on my doghouse, but for a large home unless you have zillions, I recommend you go to your bank for a construction loan and have them help you on the payouts. If you have some money and are comfortable with finances but not rich, do get a construction loan.

By having a construction loan, in the event of an emergency, your loan is already there and you won't need to stop building to take care of this problem. Work with your lender on a construction loan. The cost is small in comparison to the protection and convenience you get.

I think being your own contractor is terrific! It's certainly a challenge. It will take a lot of time, there will be headaches (and mistakes), but when it's done, the feeling you get is of accomplishment and pride. I also think, by being so close to what you've done, if anything ever goes wrong, you'll be able to go right to the problem.

It's like the *organized disorder* I have in my garage with tools. Regardless of where I put them and no matter how jumbled up or in complete disarray they may seem, they are where I put them and I can find them instantly. Now, if one single tool is borrowed and put back an inch from where I, personally, laid it, I'll have to hunt. Hasn't this happened to you before?

The same holds true for my office. I am basically neat and organized but sometimes, in a rush, I might toss something down. After a month of tossing these things, it looks like a mess, but if nobody else touches that "mess" I can find a two-inch scrap of paper in a ten-foot pile of papers or a bent thumbtack in the far corner of a deep drawer in a heartbeat. This happens when you build something on your own.

So, those of you who dare risk job and marriage to contract your own home, I say . . . well, I say **decide for yourself!** The main element is **time!** If you have or can find the **time** to do it correctly, I say there's no greater rush than reaching the top of Mount Everest.

YOU

The last individual on this list is **you,** the homeowner.

Your total responsibility is giving the final okay to things and—paying the bills! You pay the engineer and/or the architect for their plans and agree on a fee for the contractor.

You're also the coordinator of the next person involved; your banker. They finance your home because you are a good risk. Choose carefully on banks and bankers. They can be one of your biggest problems or your greatest asset. Your banker works with the contractor and you.

Many homeowners feel that if they build on their own, that they can build the perfect house. The fact is, you're almost right. When you build your own house you'll build things exactly the way you want, and regardless of how nonsensical it might seem to anyone else, it's your house. You might, in reality, build the near-perfect house; rarely the perfect house. Because the odds are heavy that you'll make a few errors. So what? As I said earlier, you can do it! I say do it and enjoy yourself.

My advice to you is if you want a perfect house, hire God! If you want a good house, hire a good contractor. And if you want a house you feel great about, do it yourself. But the perfect house, only He can build.

Chapter 4

PLANS AND SPECIFICATIONS

Let's begin with something rather simple such as remodeling your kitchen or bathroom. Regardless of how easy it seems, you will need plans! If you're putting in a shower stall or hiking up the shower head so it will wet from the top and not the middle of your body, you know, where you won't have to stoop and scoop under the shower to get your head wet, no big deal. Just measure, punch a few holes and then cover the hole you made (that was too big) with an oversized washer.

But, if you're adding a complete room or a second floor or a master suite, even remodeling your kitchen or bath, certainly when you're **building a house,** you need plans and specifications (specs). Don't even attempt doing it without them. The mess you make will be far, far more than the cost of doing it correctly the first time.

Here's a list of what I call a good set of plans and specs and I'll detail the importance of each.

SITE/ROOF PLAN

A shot of your entire building on your site. It will have all the requirements listed, such as building setbacks, utility easements, tree placement and other pertinent information that deals with your property. It should include your roof layout showing gables, hips, ridges, and valleys. It will also show your driveways and sidewalks. Figure (a) shows a typical site/roof plan on page 57 and also the Wall Sections. This the exact way plans are drawn.

FLOOR PLANS

Floor plans tell more than any of the other drawings. They show the location and dimension of each room, wall, window and door, commode placement, and closets. It tells about finishes on the floors, walls and ceilings, types and sizes of windows and doors to be used and where. Figure (b) page 58, is a typical floor plan.

ELECTRICAL PLANS

In many instances the floor plan will have the electrical plan on it also; wall switches, receptacles, lights, wall fixtures, everything electrical and their location. In some cases the electrical plan and floor plan will be separate. I like them to be separate because it makes it easier for everyone to read and your framer doesn't care or need to know anything about your electrical setup. Your electrician, on the other hand, works with the framing. Figure (c), on page 59, shows the (separate) electrical plan.

EXTERIOR ELEVATIONS

Basically, it is a two-dimensional drawing of all the elevations of the house to give you an idea on what it will look like. It doesn't tell a lot other than the floor lines and how high they are from the ground. Figure (d) and (e), on pages 60 and 61, show front, rear, right elevations.

FOUNDATION PLANS

These show how your foundation is put together; about the types of steel and concrete and where they go, how deep to dig and how much material to use. The two drawings your foundation sub contractor will use are the foundation plans and the site plans.

These plans are drawn by the architect and designed and sealed by the engineer. Since this is the first stage of building, it is very important. Should a mistake be made, a foundation repair means a monumental chore. **Figure (f) shows the foundation plan on page 62.**

WALL SECTIONS

Small drawings from the ground through the roof on how the building is put together, how the wall sets on the foundation, what the wall is made out of, type of insulation in the wall, and how the roof meets the walls.

It also shows you how the wall is connected to the foundation, if you have hurricane straps, the types of sheathing, the type of insulation, and whether it's brick, block, wood siding or logs. **See Figure (g), page 63.**

INTERIOR ELEVATIONS

The more detailed drawings are of bookshelves or cabinets, mantels, and vanities. Do you have drawers or doors? Which way do the doors open? Built-in desks, wet bar, entertainment centers, what? All of these things are drawn to a one-inch scale on the plans.

These go to the cabinet maker or interior trim worker who then do their own drawings—in more detail. These are called "shop drawings" that they make but they must have an idea on what is to go where and give the dimensions so they can bid on the job. The drawings show it all. **Figure (h) show interior elevations on page 64.**

The drawings on the next page are those of the Site/Roof Plan that also include the plan for the Wall Sections. I use this drawing twice because it's the way they come and to split them up tends to be confusing. Confused yet?

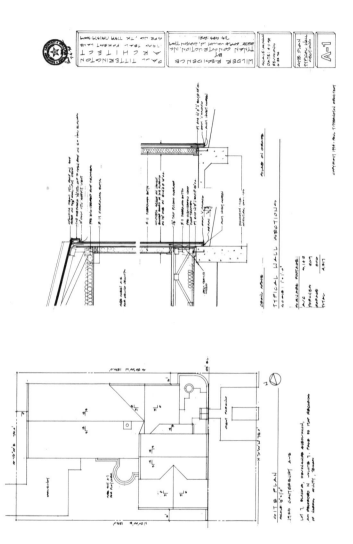

Figure (a) [Left side only] is the Site/Roof Plan.
The other is the Wall Section, shown again on page 63

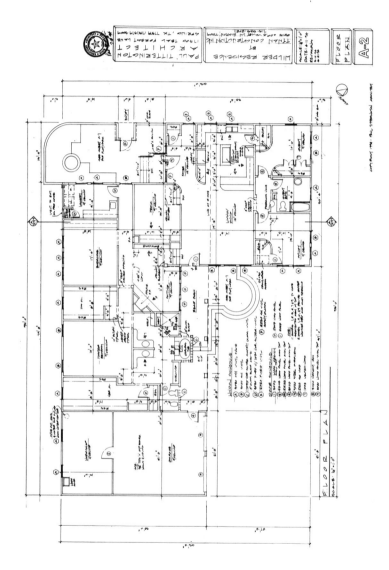

Figure (b) Floor Plans.

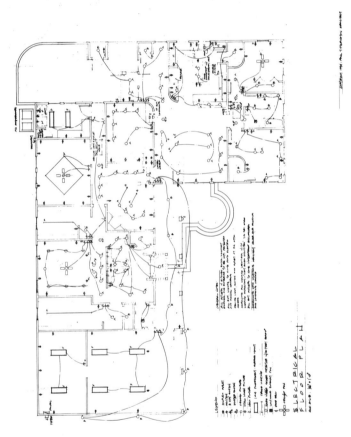

Figure (c) Electrical Plans.

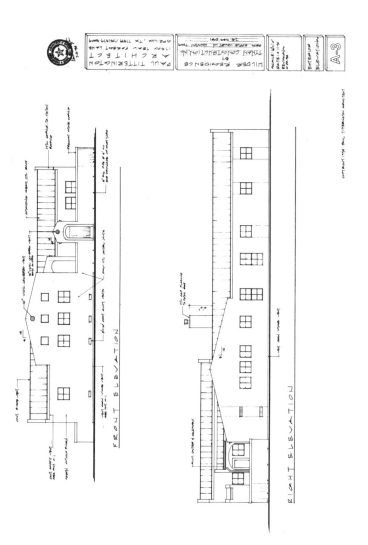

Figure (d) Front and Right Exterior Elevations.

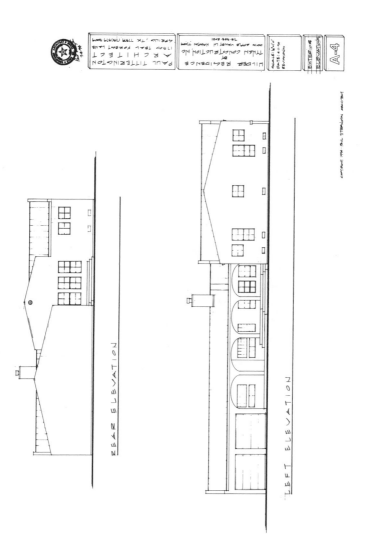

Figure (e) Rear and Left Exterior Elevations.

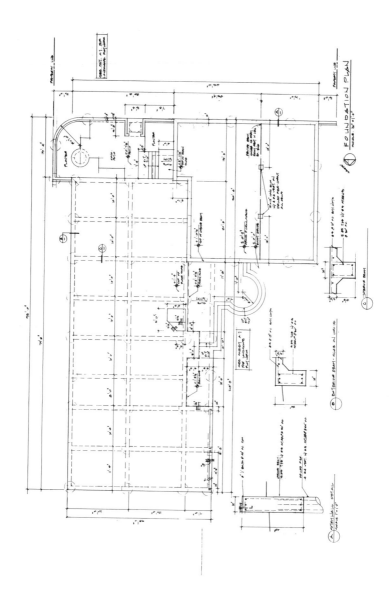

Figure (f) Foundation Plans.

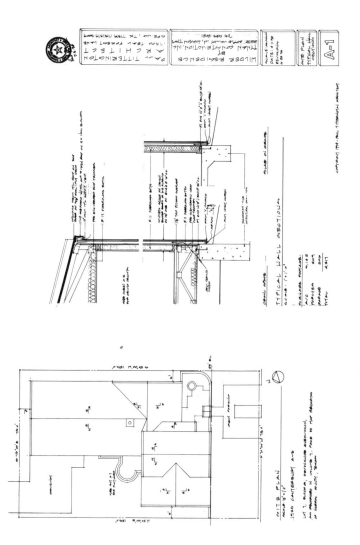

Figure (g), the Wall Sections are on the right.
(Notice, the Site/Roof and Wall Section Plans are together)

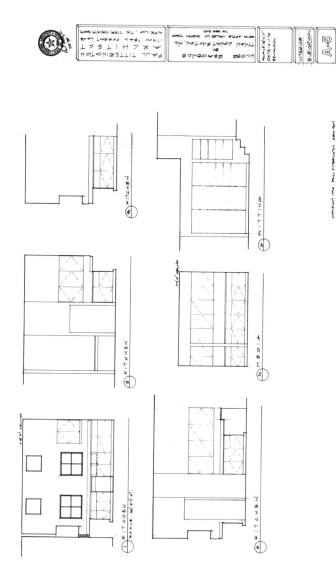

Figure (h) Interior Elevations.

OTHER DETAILS

There are so many details to tend to, even in installing a simple mirror. A contractor might install a 2 x 4 foot mirror in your bath whereas the plans call for a 4 x 8 that covers the entire counter top, or a smaller skylight because the smaller ones were on sale. They might be trying to save money for you or get the job done faster but this isn't what you want.

By now, I know, you've put this book down and I've convinced you to find a house you think you'd like that is already built. Or maybe I've convinced you that you can't do this yourself and you are rushing out to find a contractor because it does, in fact, take a lot of work and planning to build or remodel.

But you'll need to read the rest of this book in order to sound kind of bright to your contractor. Then, I promise, you'll be able to double-check things and know some questions to ask.

If you plan to have a large fireplace, what's it going to look like? Will it be brick, marble, will you have a big mantle or hang a moose head over it? Will it be 1 inch off the floor, 24 inches off the floor or will it be flush with the floor? Will you have a hearth or will it run straight down? Will it be gas or log or with a log lighter. Your interior elevation tells all of this.

All the trim around kitchen and bathroom cabinets need to be listed. Another important item is what is now called a media room. They have big screen tv's and stereo or CD cabinets and these have to be drawn to specs.

If you're going to act as your own contractor, one of the first things you need to learn is how to read plans. Looking at a set of plans for the first time is a frightening experience. There are symbols, lines, drawings, numbers and so much that won't make sense to you if you don't know what they mean.

Look over these plans in this book. It will serve as a starting point of what to ask for when you go to buy a set of plans. When you get your plans, ask the architect to explain them. It's part of their responsibility and, trust the fact, this advice is included in their fee.

Remember, it's not only *okay* to ask your architect questions, it's wise. If there is anything you don't understand, please ask! Architects love to answer questions, especially if you, as a first-time builder, know what you like. Your architect will tell you if it can be done and how much it will cost to do it. Ask!

Chapter 5

HOW TO SET A BUDGET

Budgets are the heart of a job and are extremely important; it's how you can track a job or decide to even start one. Begin with something simple like remodeling your kitchen. The first thing to contend with is demolition; before you rebuild something you have to tear it apart.

PAD AND PENCIL

But even before that, get out a sharp pencil and a legal pad with several pages and list everything you need to do. Make a list of the cabinets you are going to refinish or replace, what kind of floor will you have, will you keep all the appliances or just a few, the type of light fixture you plan on getting, wallpaper, tile or paint.

Measure your floor, wall and counter top and record these measurements. Go to a local Home Improvement Center and price the fixtures, appliances, cost of linoleum or tile, wallpaper, a new counter top, new faucets, sinks, new designer cabinets, etc.

USE THE TELEPHONE

Then, call a few companies who do wallpaper, tile and counter tops and see what their charge will be. The total cost might frighten you to a point that whatever you have now might not be so bad. Most homeowners don't need to remodel their entire kitchen (many won't once they tabulate the cost), perhaps just a few areas will make all the difference in the world.

I have a friend who has a home that is maybe 15 years old and his wife wants their kitchen remodeled. This doesn't mean tearing the entire thing out and putting in new things. Over a period of years when one appliance broke and was replaced by another, the couple, both working and on a limited budget, replaced appliances when they got a deal. Everything works, but it looks awful. The refrigerator was beige, the dishwasher was white and the stove, stove top and wall ovens were yellow.

They priced a new stove and stove top and found they could afford that. Since their refrigerator was newest, they chose a beige color to match. They then found out that they could get "fronts" only for their dishwasher from the appliance store and with a few screws and a few dollars, they got beige plates for it.

When they priced the wall oven, it was about $1200 and was out of their budget. The oven worked but the color stood out like a . . . well like a yellow oven in the midst of beige appliances. It took everything away from what had been done. So, they opted to get the oven doors *baked* to match the other appliances and the cost was only $100.

Next, they replaced their out-dated light fixture with a new fluorescent fixture. You'd be amazed how well the changes made everything look. But they glanced down at the floor and realized it needed to have something done to it.

They measured, called a tile contractor for an over-the-phone ballpark bid, took their measurements to a building center and again, they could make a decision.

They found that tile was too expensive. In looking more they opted for a linoleum tile that looked like real tile but was far cheaper. Their kitchen is gorgeous. The counter tops are not exactly what they prefer but not offensive. The sink and fixtures are fine, the paint and the wallpaper on one wall needs to be replaced. They're doing that next.

But let's say you have the money and the time and feel the need to do a complete remodeling job on your kitchen. Besides, you're planning on selling your home soon and you know a major selling point is a remodeled kitchen. Now comes another choice. Do you . . .

REMODEL COMPLETELY OR PARTIALLY

Will this new kitchen increase the value on your home or will it just make it easier to sell? Let's talk about that for a minute. You bought it, say ten years ago, for a certain price and it's now worth more. Will the cost of remodeling the kitchen be more than the return on the investment? These are judgment calls.

If you're going to definitely sell your home, just put in something neat and new-looking, not especially to suit your taste. Tastes differ. If you plan to stay if you don't sell the house, you have to then make the choice to do it your way and to determine how much it will cost to do it this way.

The ideal justification for putting remodeling money into a home is when you have the funds and you just want to make changes that please you. I enjoy doing work for people like that because I can go in and do the job, I can install what they want and make them happy.

So, whatever your situation, if you plan to remodel make that list of what you want and add up the total. You have to lay out every step of the job. This makes sense, doesn't it?

If it is your kitchen you are remodeling, don't forget new swinging doors from the kitchen to the dining room? Will you put in a breakfast area? Is there room? Will you have to take some out of the dining room? Will you need a new plan for enlarging the pantry; shelves that turn around like a Lazy Susan? A new fridge with ice maker and pipes that have to run from your sink to wherever the refrigerator is placed? Do you want gas and not electrical appliances? Will the new sink meet the counter top? Should you tile the counter top completely or just around the sink? What about new windows? And the list goes on and on.

There's so much to consider and the most incidental items add up. These costs must all be taken into consideration before most can make a decision. And be careful not to add too much space or you'll be tearing out more walls for more major reconstruction.

ADDING A SWIMMING POOL

Even with adding something major like a new swimming pool, you have to determine if the cost will increase the value or salability of your home or do you just want it! Swimming pools used to be luxury items in the past that everybody wanted and had to have whether they could afford them or not. I've seen neighborhoods where every home had a pool. I think it was a "Keeping up with the Jones'" thing.

Know that if you put in a pool it might not increase the value of your home one cent! In fact, many people don't want a pool because of the danger to small kids, because of the liability with neighborhood kids, the cost of maintenance or they might just want a nice, open yard. It really depends on your neighborhood and your personal situation or preference.

On most small jobs (once you've read this book) you can act as your own contractor; it's not like building a house with all the demands on your time. You can oversee this job by taking only one week from your vacation, and once you decide to go ahead with your remodeling, you can call individuals to do the work and you buy the materials to save money. Or, you get one contractor to do it all. Less headache but, of course, more cost. If you've decided to act as your own contractor on this one, here's the procedure.

CUT SHEET INFORMATION

You have to get what we call a "cut sheet". At the end of this chapter you'll find an example of it,. In remodeling, nothing goes as smoothly as you'd like. There are always surprises and each surprise costs money. When you tear something down there will be problems that weren't added in when you just looked at your list from outside the walls.

This cut sheet allows you to track your job from your guesstimate to your actual costs. If you're short on funds, you might have to make a cut; to get maybe a less expensive light fixture or another type of cabinet because a pipe in your wall is connected to another that all needs to be replaced; or for the space the new cabinets you want takes up so much room, it becomes either those cabinets or a foot-wide refrigerator. Yes, you'll have to make revisions as you go along; plan on expecting the unexpected because it, invariably, happens.

DECISIONS—DECISIONS—DECISIONS

These are the three decisions to be made as I see it: First, how much money do you want (or have) to spend? Let's say you're going to do major remodeling, like a new kitchen, maybe both bathrooms and perhaps adding a game room and you figure the cost to be $30,000. This is getting out-of-hand already, isn't it? You have only $10,000 to spend then and you'd probably like to do it for even less than that!

But let's say you are going to do a big job and you have the $30,000 to spend. I suggest you take 20% from that figure and see if you can build your basic project for the remainder, for $24,000. The reason is to cover the surprises you'll run into plus you will want to upgrade certain things as you get farther along in the project; everybody does! That 20% is your cushion.

I tell people this whenever I build. If they plan to build a new home for $200,000, I ask that they use this 20%-off formula and try to build for $160,000. You will need that extra 20% or $40,000 cushion to do things you didn't realize you wanted when you began. And if I'm wrong, you're that much farther ahead. You have the new house and that cushion-money to do other things with.

Second, if you are going through a bank or lending institution for this remodeling money, ask them for the $30,000 and still try to build for $24,000. Now, you've set a really good budget. You're smart, you're ahead of the game, you've done all your homework and you're ready to begin.

Do you see what I'm doing? I'm trying to make this building and remodeling easier for you and to avoid a hassle with labor or material, with running low on funds from your bank, if you plan and see a budget, well, it's the only way to do it.

Third,get the entire family involved. See what your son wants in his room or your daughter in hers. Find out what makes your better-half happy and what pleases you. Talk it over while having dinner, carry the conversation into your den, shut off your TV, unplug your telephone or put on the recorder and talk.

Make these decisions as if you are about to go to war. Determine what you want, what you need, what you can do without and what you just might have to accept. If you get her room and his room and your room set, then go into the Big

Room or Den or whatever it's called these days, the Great Room I think is the new terminology.

This Great Room is important because it has shelves for library books, maybe a cabinet for the TV, a bar perhaps, tiled floor, hardwood or carpet. Will it have a skylight, will there be a vaulted ceiling, how about a back door or sliding glass doors to the patio?

This is the focal point of your house. Young people could care less about a dining room or even a living room; they live in the Great Room as do most of the family and it's the room for visiting friends and relatives, for eating and watching television, for family conferences and for relaxation.

Yes, plan, discuss, change plans and do it according to budget whether building or remodeling. Put down on paper what you want to do, measure, price, get bids and use my 20% off rule so you'll have money for a vacation or for the extras you didn't count on.

I know one man who has been building his house for three years and he's $200,000 over budget because he never set one! I advised him that what he wanted was unrealistic to begin with but he thought he could do it and he has failed. Don't you make that error. Set a budget!

There's one other thing I need to comment on with a budget. As you are following your cut sheet through the job with foundation, framing, plumbing and with fluctuating lumber prices that go up and down and they do go down, believe it or not, and you are always readjusting and compromising. Don't compromise on the important parts like framing, foundation or roofing; you do this on carpets, counter tops and fixtures.

YOU CAN'T DO IT ALL TODAY

Bypass that $3000 imported crystal chandelier in the entry for now. Get a $300 one and pass the inspection. A year or two down the road when you've made or saved some more money and you still want the chandelier, buy it then.

Compromise on the things that become outdated or wear down. Carpet for instance, wears down quickly. Compromise on it rather than your framing, foundation, roof or GFI's for your electricity. Don't compromise on your plumbing or air conditioning. Do compromise on color of paint if it is too expensive. Paint the entire house white and go back next year and paint the color into the rooms one at a time.

Light fixtures (like the high-priced chandelier) can be compromised on also. People have no idea how expensive light fixtures can be. I know you'd like to have the $200 fans throughout the house, well either don't put these fans in at all or get a $40 or $50 fan for each room; they're everywhere!

This is my objection to many builders who build tract homes; they put in beautiful microwave ovens and crown moldings and expensive fancy fixtures and try to make the cost reasonable. If so, they have to cut somewhere and it is usually in the framing and foundation and within 8 or 10 years, your beautiful microwave is sitting on a buckled floor or cracked slab and the roof has caved in breaking that fancy molding and ruining those expensive fixtures. It's foo- foo with no substance and I go to war against such building practices.

I'm not trying to make you unhappy with this or make you wary of buying a tract home, but I want you to be realistic. You have to pay for quality and when you try to fudge on the wrong things, it costs more.

Be happy with what you can do and compromise on the things that really won't make much of a difference and get them later. Be easy on yourself. Set that budget and be smart and get what you want as you move along and make it more fun that way. You know, little "happys" now and then to upgrade what you've already done.

IT'S A MESS

I have some neighbors, people I love dearly, and they are remodeling their home. *She* wants it all done now! *He* wants to play golf on at least some of his time off and maybe take the family out to dinner.

They have set somewhat of a budget but the somewhat isn't cutting it. They plan on doing things, then when they get into it they find that their appraisals of cost are seriously inaccurate. They have run out of money and had to put things on hold and they, without a plan, have almost every part of their home half finished!

I've made suggestions, even comments, but they continue trying to do it all at the same time instead of concentrating on one area, resting and relaxing in between, then starting another spot. I don't know if it will ever end because, I promise, by the time they finish, there will be more changes and upgrading to do.

First, they took off an upstairs balcony that was adjacent to the fireplace. An "unexpected" came up in that the fireplace had loose brick outside. Upon further inspection, they discovered they also had loose firebrick inside. The house is over ten years old and the outside brick had to be matched. Well, they found some brick that "almost" matched. (Some of the 20%)

They then decided that they didn't want to repaint their house; not now and not ever again so (woe is me) they went for the vinyl siding. It's taken them, both working and getting whatever friends they could to help, a few months to do the siding.

But wait! The siding job isn't finished yet and she decided they had to remodel the upstairs bath, and make one large room out of two. In the meantime, why not change the entry way into the living room? They're doing most of it themselves.

The outside of the house is maybe 80% complete,

some siding and some gray unpainted boards. Their living room is in disarray because the new entry way can't be completed until the sheetrock man tapes and floats the upstairs remodeling work and that needs to be done at the same time to cut cost.

In the meantime, the new furniture has been delivered and is in the dining room, some of the other bedroom "stuff" is in the living room, the stairway is cluttered with mess from tracking outside in and up to where the repairs are going to be made, the guy hasn't played golf but a half-day in the last three weeks and he was too tired to swing a club the day he did play.

His and her vacations are almost over and then they'll have to work weekends and/or nights and well, it is a product of poor planning and wanting to get everything done right away. Like I said, I love 'em but I don't understand the necessity of having to do it all now.

In building and remodeling, there truly is but one way to these things correctly and I'm listing them all in this book. Do you see what I've done to you? I made it so you just have to read my entire book if you want to do it correctly.

SAMPLE CUT SHEET

Job Description	Labor Cost	Material Cost	Estimated Cost	Actual Cost

Chapter 6

PERMITS - INSPECTIONS - CODES

Every municipality in the country has a different set of rules. The two basic codes for building construction in the United States are, the Uniform Building Code and the other is called the Standard Building Code.

The Standard Building code is also known as the Southern Building Code. The main difference is that the Uniform Building Code regulates about 80% of the country whereas the Standard Building Code (Southern) regulates coastal areas with special provisions for storm and hurricane building requirements.

For places like Florida, the Texas Coast or parts of Louisiana, this code book doesn't tell all because each area makes **addendums** to the code. All cities have these addendums that cover certain problem areas that needs to be addressed.

The city of Miami and Coral Gables, each located in Dade County, has a separate set of addendums. To be clear on what they are, call the city that governs the building codes and building requirements in your area.

You can purchase a copy of these addendums from your city building department. Then, just snap them in your basic code book. Building bookstores will have the code books only, no addendums.

You'll have special requirements for septic tanks, water wells, building construction and flood-prone areas. Some are enforced and some are not. For instance, in Harris County, Texas enforces codes for water wells and septic tanks but not for construction.

PULLING A PERMIT

To find out what drawings are required to pull a permit, you need to either call or go in person to the building department. They will furnish a list of exactly what these requirements are. Some of the things you might need to furnish are soil reports, land surveys, engineered foundation plans as well as a full set of plans and specifications.

When you're ready to pull your permit, I suggest you take along at least $400 cash with you to the building department. Most will not take checks or credit cards, not even a cashiers check. Bring cash. If you're using an architect or contractor, they can do this for you and they know who to go to, when? how? why? etc.

In many cities there are people who do nothing but pull permits for builders and you can find out who they are by asking at your local building department. These people are familiar with the procedures and most of the time they are friends with the various inspectors. They can accomplish in a few hours what it might take you a few days to do. Once the plans are approved, they will be stamped in red, APPROVED!.

At this point they issue a building permit and a copy,

one to post on your job and the other to put in your files in the event the posted permit gets rained on, torn, stolen by an irate sub or whatever. During every inspection, you'll need a approved set of plans available for the inspector. Most permits have a one-year time limit.

Next, you have to follow the code requirements and get ready for inspections. Usually, the foundation inspection is first. Before you pour concrete you dig holes, put in plastic sheeting and steel but before you pour, these inspectors want to come and see it. They want to make certain that the type, the amount, and the gauge of steel you're supposed to have is what you do have.

In most cases you order this inspection the day before you pour your concrete. it's necessary that you are there when they are scheduled to inspect. It can be done the day before. Schedule it for the morning or early afternoon or early evening. If you're not there and there is a problem, they "red tag" the job and you have to reschedule the inspection.

In the meantime, the concrete is on the way and cannot be poured. If you are there chances are you can remedy the problem with a few workmen and proceed as planned. It is inconsiderate as well as pretty darn dumb not to be there when an inspector arrives.

These inspectors know the problems. Even if they have an objection and you're there to tell them you'll fix it, chances are they'll issue you the green "go ahead" tag and leave, trusting you'll do as you say. That is, of course, if the problem is minor. On a major problem, whether you're there or not and if it isn't fixed, expect the red tag.

Inspections come in sequences such as: First, you get your building permit from the city along with your building plans. There is a number on these plans and that number is given to your plumber, electrician and air conditioning sub. Each of these people must then go to the city to pull their permit that relates to this particular job.

Now, you're ready to *break ground*, builders language for beginning your project. Now comes the parade of inspectors. The first one is your. . .

UNDERGROUND PLUMBING INSPECTION

The plumber handles this part but you should oversee it. The plumber puts in the plumbing then calls the inspector. If the inspection passes, you should then pay your plumber his first draw (money). If it doesn't pass, the job is delayed until the plumber gets whatever needs to be fixed, and the inspector again comes out.

The first inspection for the contractor is the foundation. The plumber is solely responsible for his inspection, as is the electrician and your HVAC (heating, ventilation and air conditioning) contractor.

FOUNDATION INSPECTION

The Foundation Inspector likes to look at the steel before the concrete is poured. Some inspectors even like to be there when the concrete is being poured to make certain no water is added and that it's the right mixture.

You follow the same procedure as you did above; get your green tag, pour the concrete, pay the foundation people and then things get a little easy for a while.

FRAMING INSPECTION

Before the insulation and sheetrock goes in and before the walls are up closed up, these inspectors want to look at your framing to make certain it meets all code requirements. This is also one that the contractor is responsible for.

HVAC ROUGH-IN INSPECTION

They want to see the duct work, refrigerant, gas lines and placement of the air handler and furnace. Also, exhaust fans for the bathrooms and kitchen.

PLUMBING TOP-OUT

This is when they inspect the supply lines and drain lines in the walls before the walls are closed up. Remember again, these inspections and inspectors, as much of an inconvenience it may seem, they are for your protection! Be happy they are there.

ELECTRICAL ROUGH-IN

They check the wiring throughout the house that is inside the walls and running through the attic. The inspector won't climb in the attic; they expect to see the wiring from floor level. Each is responsible for their individual inspections. If one is faulty they will hold up the entire job. After each inspection is given the go-ahead, the subs are to be paid.

FINAL INSPECTION

This is when they inspect everything! The electrical part means all the fixtures, switches and plugs have to be in place. All your appliances are to be hooked up, no loose wires. When this is passed, you can have the power company hook up to the house.

Your HVAC contractor needs power to the house to set the final condensing unit, to charge it up and make certain it runs properly. His work is subject to a final inspection on this also. And now, your plumber has to set the sinks, commode, lavatories, tub, and water heater for his final inspection.

One inspector does a final walk-thru and if everything passes, they will issue a *Certificate of Occupancy*. Until this certificate is issued, the homeowner cannot move into the house. Sometimes, in some areas, you must even have most of your landscaping done prior to a Certificate of Occupancy being issued.

Let me share with you a bit more information on inspectors. Some of this, I realize, I'd said before and I might even say it again and it's not because I forgot, it's because I don't want you to forget!

HANDLING INSPECTORS

1.   Call them as your schedule states (usually
      the day before).

2.   Be there when your inspector arrives.

3.   Treat each inspector with dignity. Be cordial,
      (be smart) and offer them a cup of coffee
      and/or a donut.

4.   Ask their advice. Walk around the project
      with them and ask questions. They know their
      jobs and are knowledgeable and might tell you
      something you didn't know that will help with
      other inspections.

5.   Don't ever—not ever—be a smart aleck or
      make an inspector angry. They will write up
      a list of complaints you never dreamed existed
      and then leave. Then, another inspector comes
      out and chances are strong that they will find
      other things wrong. Courtesy is the key word.

I'll share a story with you about an inspector who was
on one of my jobs, not a bad guy, just a different guy.

The city was small so this one inspector's job was to
go over everything. The problem was that his former
profession was as a licensed plumber. He was nice enough
but he was dedicated to making the world a better place and

the only way to do that, in his opinion, was with flawless plumbing!

He checked each pipe, fitting, joint and drain. He tested knobs, handles and shower heads. He measured and leveled and paced distances and he made certain that this job was plumbed correctly to the finest detail.

The problem was, he didn't care that much about anything else! He walked through the jobsite like he had lost his tube of *Preparation H* and was late for an appointment with his Proctologist. I was passed on everything immediately!

We got along well, I stroked him whenever I could without being obsequious (good word, huh? I looked it up. For those of you who, like me, didn't know the word, it means fawning, polite to a fault, almost like kissing his feet. I saw it and decided to put it somewhere in this book).

The problem is, most contractors know that a **good inspection** is like having a partner, another expert going over what has been done. They want a fair inspection and they want a complete, good inspection. If something is wrong, they'd much prefer fixing it then and not having to come back a few weeks or months later.

Thank goodness, on this particular job, nothing went wrong. It's been a few years since I built the house and no complaints yet. Passing the inspection should be only when whatever is being inspected is checked out! Make no bones about it, these inspectors are qualified and efficient and only on rare occasions do I get an inspector who is hurried, who is not complete, and/or who has a nasty personality.

I have never experienced anything such as an inspector asking for money or favors, but I have had a few inspectors *Red Tag* me to the point that it was absurd. In either case, call the downtown office, talk with their supervisor, tell them you're not being treated right and you'd like to request that either another inspector or a supervisor give their appraisal.

My experience is that supervisors hate to leave the comfort of their padded desk chairs, air conditioned office and regular coffee breaks. Most go to work dressed with a shirt and tie and prefer not to get out in that hot sun; they did that as inspectors. I have several friends who are supervisors and they might not like my saying this but all of them (hopefully) have a fun sense of humor and I know they'll agree and laugh.

These supervisors have "been there" and know each inspector. The first thing the supervisor will do is to ask to speak to the inspector over the telephone. They discuss the problem and try to work things out. If it can't be settled, they usually send another inspector.

Since most building goes on in dry weather, usually the summer, it gets awfully hot on these job sites. Of course the supervisor wants to work it out over the telephone if at all possible. As a former carpenter and now a contractor, I understand the supervisor's reluctance to leave his air conditioned comfort.

Some municipalities require an inspection whenever you pour concrete; for walks, driveways, patios, etc. To be safe, whenever you build, call the Building Office and find out if a permit is required.

Chapter 7

BEING OR HIRING A CONTRACTOR

On any sizable home improvement project, this question is foremost in every homeowners mind. In the words of Shakespeare when he was deciding to build his own home or hire a contractor to do it for him, *"To be or not to be (my own contractor). That, is the question?"* His decision was, "Should I just call various workmen and have them do the job? All I have to do is tell them what I want and supervise a little and it's done. Why pay a contractor to do this? It seems relatively easy to me and I can save a ton of money and write a few more plays?"

ARE YOU LEGAL

The first thing to know is if it's **legal** for you to do your own work. In Florida for example, it is extremely difficult to obtain a permit because of the intense code requirements concerning licensing for contractors.

To do your own contracting in Florida, you have management responsibilities, such as liability insurance, workmen's compensation insurance, code requirements, etc. In Texas though, you have a free hand. There is **no licensing procedure for the people to build.**

Every state is different. Find out the local requirements for building and obtaining permits. Do this simply by calling the **Building Inspectors office** in your city. Anyone in that department should know or they can direct you.

DO YOU HAVE THE TIME

After you obtain the necessary permits, you must then determine how much time you have to put into this job. It may look easy to you but you have to be there to supervise and take responsibility for the workmen. You can't be there only 20% of the time to look over 100% of the job. If you have a regular job or a family, I'd say pass it up and hire a contractor.

I'm not trying to discourage you on this, it can be done. All I want to do is point out the facts so you can make this decision with a degree of knowledge.

One high-priority responsibility is to make certain everyone who is supposed to work, is working, and that they are following your instructions. If someone is starting to do something other than the way you want it done, you had better be there to make that correction immediately. The longer it continues unsupervised and uncorrected, the more chance for error, and error = added cost.

You will have to be there to order extra materials too, because no matter how exact you think you are with measurements or quantities, something will break and has to be replaced or you'll use more of this or less of that. I suggest that you, personally, write the checks for additional materials or supplies.

By pointing out these various pitfalls of doing it

yourself, get a pencil and paper and mark down all that it entails. Then look at it realistically and if you feel you can, look at it again and make your decision. The time it takes, I've found, is the major consideration of self-contracting.

The minimum time you must be there is in the morning to make certain everyone gets started, I'd say like an hour or two before you go to your regular job. During your lunch hour, at least call and see what additional materials are needed or what has to be done. After you get off work, go back to the job and do a visual inspection of everything that was done. If there are problems, you have to make a long, detailed list of each one. The following morning, talk to the person who is in charge and get them to correct these problems.

It is a lot of work and a big responsibility. Wouldn't it be better for you to just earn money the way you know how and leave the building or remodeling up to a contractor who does that for a living? Sounds like I'm trying to discourage you, doesn't it? Not so!

Those of you who know me or have heard me over the radio or watched my tv program know that *"I march to my own music."* I say things that I feel are right. If I'm for or against something I tell you about it and I tell you why. My intent is not to hurt any manufacturer, supplier, or retail outlet. I pride myself in being academically fair to all products or businesses. If I think it's bad, I tell you about it even if a relative is selling it. And if I feel it's good, I tell you that too. I deal with performance and fact.

I have over 15 years of experience as a contractor and

I know the problems. I realize there are those who have 40 or 50 years experience but there's little that is new; just the same problems and same types of people to deal with over and over. And, I talk to you as a friend and as an advisor, not as a contractor or salesperson.

HOW MUCH WILL YOU SAVE

As you read through this chapter you can determine for yourself whether it saves you enough money to warrant your going through this headache. And it is, in fact, a headache-producing job. Let's determine, now, how much money you'll actually save. How much do contractors make?

The consensus of opinion from many is that men or women who build homes are going to get rich off of you. This is not so! Most small contractors are living modestly, some close to starving, borrowing money, running in the red most of the time, no cash flow. If a contractor can gross 20% on every house they build, it would be perfect. It costs a substantial amount of money to be in business.

There are office transportation , insurance expenses and advertising. You have permanent employees, you have to have money in the bank for materials and cash flow for suppliers. The reality is that most contractors end up earning 8% to 12% on a job.

Let's say you plan on building a home for $200,000 and you feel that you can save $40,000 by doing it yourself. That is totally false! You might end up saving 15 to 20 thousand dollars and the errors and headaches, not counting the time you put in, is simply not worth it! My experience has been that the vast majority of people who built a house that was to cost $100,000 or more have actually lost more than they think they've saved because of error.

This is how I see it and I speak from experience. If you plan to build a small house or a beach cottage or cabin in the woods and you have time to play with it and you enjoy doing it yourself, make it a family thing. Get your family to camp out on the site on weekends or during the summer and do things slowly and do them together.

If you have the time to build on your own and it's not a complicated project, I say do it. Rent or buy a small trailer to camp out in and make it easier . . . and fun. Do it, save money and enjoy it!

If you have the bucks to build a $200,000 house or more, get somebody else to do it. You can still visit and supervise but at your own choosing. I recommend you get a contractor to build anything other than the camp or cottage I mentioned above or a small remodeling job around your house.

DO YOU HAVE ORGANIZATIONAL SKILLS

Contractors are like orchestra leaders. Sure, you might know about carpenter work but are you a "rough" or a "finished" carpenter? Some carpenters do framing and others do cabinets. And, how much do you know about plumbing, electrical, brick, tile? Are you familiar with the current codes on safety, the new material, the various windows and doors?

How about financing; bank loans, dealing with a mortgage or savings and loan company, some lending institution? And, you'll have to drive to City Hall to pull permits, begin arranging your time to coincide with the various inspectors time, you know, the guys from the city or county that come out to approve or disapprove the work that's been done?

The ways you can save (or spend) money are limitless in building. Do you nail those two boards together or throw them away and get a new board? Are you planning to put my famous continuous ridge vents on the roof or can you get away with turbines?

You ask yourself, "Should I use one-quarter inch plywood or that more expensive three-quarter inch? I already have the lightweight stuff. I wonder if two by fours can go 36 inches apart instead of 24? And, I can forget these GFI's can't I? The wiring is put in according to code?" These questions and many, many more need to be answered to help you decide.

There are thousands of decisions to be made and they have to be made quickly and logically. If not, something caves in and you must go back and do it again, this time correctly. Try to cut corners and it will only cost more. And the time factor is monumental. Maybe they did it in a day but it will take you a week!

Whew! I'm getting weary and I know I'll have nightmares when this chapter is over because it reminds me of my contracting days. These guys earn their keep. I say pay 'em, be nice to them and smile at what they are doing. A good contractor is an orchestra leader and in the front row of that orchestra are the banks, permit offices, and inspectors. Second row are the foundation people and plumbers and right after that are the framers and roofers and, oh, the list goes on and on.

You see, if you contract your own job, you have to deal with individual band members (so to speak) that were not part of an organized band. The plumbers are aware of their own areas and certainly know *something* about framing and doors and windows, but they put in their stuff and move on whereas a contractor, that orchestra leader, plays every "instrument" in that entire group and can see problems before they arise and correct them before any or too much work is done incorrectly.

So, do you have the skills to be your own contractor? Do you have the time? Are you legal and do you want this aggravation and how much money will you actually save? Of course, sometimes, you have no choice. You might need to build and don't have the money so you have to take a chance on doing it yourself, you know, at a slow pace. Contractors can't do it this way so you become your own contractor out of necessity. Let me tell you how to do this.

YOU CAN MAKE IT HAPPEN

If your situation is one that you have no choice, here's the way it can be done and the way I've seen it done —successfully. On my way to meet with my publisher who lives in the suburbs, I have watched a house being built on 5 acres of land for the past seven months. The fact is, the guy and his wife work full time jobs and can work on their home only on weekends. Chances are they don't have the money to do it all in a few months so they are doing it themselves, working on it whenever they can and when they can pay for it.

I smiled when I saw the whole family out there hammering or moving dirt, sitting on boards or boxes eating sandwiches and drinking sun tea. I'm not certain if mom or dad was doing the directing (probably both) but will these people appreciate their new home? This is what America is all about, isn't it? I know I sound like a softie and I even had tears in my eyes when they finally, after maybe four months, started framing. Just the other day I saw where the roof was being put on and guess what? Ridge vents. I surely liked that.

Some of the times I passed and saw a few trucks and maybe five or six people working. Everything looks right and I guess this person has friends who helped, friends who were individual instrument players in that orchestra, you know, one guy who knows plumbing, one who knows electrical, roofing, etc.

But the man building the house seemed to be the orchestra leader. I saw him there all the time telling his helpers what he wanted and (no doubt) and how much it will cost to do it their way as opposed to some other way and then the decisions were made. It took this family a year or more to finish their new home and, of course, they couldn't expect a contractor to work on this time schedule.

If you are planning to build or remodel your home on

your own, do it by taking in a few classes or by reading books or by listening to radio or television shows. This book I'm writing, for instance, is just one of the items to help you make that decision. I thoroughly enjoyed being a contractor but I like telling about it more.

I found, recently, that I had some time on my hands and decided to build again. It's for a friend and a special house and special circumstances.

First, as my friend, he knows that I know what I'm doing so I will have fewer headaches. He gives me his budget, a set of plans, and I do the rest. We go over what has been done and I watch over everything.

CAN YOU HANDLE THE PRESSURE

Here's what I was told by the older contractors of what happened to the building business. Few changes or trends happen overnight; it usually takes time and a gradual process and the end result can be disastrous. I feel it's an accurate explanation of what happened.

After World War II the apprenticeship program in the building industry dropped to nothing! We started hiring unskilled labor that learned by *doing*. You never knew who you were hiring to build for you. It's that way now.

The majority of the young people doing unskilled (some skilled work) are high school dropouts who take any laboring job rather than starve. Unless you have a legitimate, experienced contractor, there will be problems. Imagine if you hired people who say they can do the job and can't.

The problems you'd have with these people would mount quickly. If you don't have experience, if you don't know what *you're* doing, if you can't spot a well-meaning worker with limited experience from a qualified one, it's the case of "the blind leading the blind".

When the inspector comes out to look at, say the framing, they don't talk to the framer, they talk to you! If you are contracting your own work, you had better have some answers. Yes, it can be scary because you're playing with a ton of money, probably more money than you will spend on yourself for a long time, maybe ever.

If you decide to tackle this job on your own, make certain you've covered all of these points. Get the knowledge and skills, have the time, prepare for headaches, think about maybe neglecting your family. You can do it (build on your own I mean, not neglect your family) if you persevere but deal with this building business academically.

Building is like climbing stairs. To get to the top you have to take it one step at a time. If you try taking two or three steps, your chances of making a mistake multiply. It's normal to be impatient but there's only one way to do it correctly; one step at a time.

Chapter 8

SOLICITING BIDS

This is yet another difficult part of the process simply because you have to deal with a lot of people who just don't know what's going on or those that don't care because this is a no fun part of this business for them. When I say this, it isn't that I'm trying to discourage you or use my own "scare tactics" to get you to act, it just happens to be the truth. Let's see why it's difficult.

First off, you have to find the sub-contractors. Where do you go? There are no associations. You won't find them in the yellow pages. Most don't have the money to advertise, so what do you do?

There are subs who might have worked for various contractors all over the state, all over the nation. Finally, the sub finds a few contractors who treat them fairly so they stay with them. These guys can be found and checked out. Most contractors have a *team* of subs they work with on a regular basis. But if you're acting as your own contractor, how do you select your own team?

FINDING SUBS

A good way to start off looking for subs is by snooping around other job sites and looking at the work being done there. This way, you can get a first hand look at things. Nobody said it was easy but if you want it to be *easier,* this preliminary reconnaissance is necessary.

Remember, one step at a time.

It wouldn't even hurt to compliment the workmen. One or many will step forth and say, "Thank you. We're proud of this work too and we take pride in our building." Even the owner of the house will not object to your looking around. He's pretty excited over what he's building and having someone interested in it is a compliment.

After having visited a particular jobsite a few times and you become friendly with the sub-contractors, pull them to the side and tell them you're planning to build. You're now striking up an alliance and it is so much easier to work with those you like and who like you.

Try to think how these sub-contractors think. You've seen their work, complimented them on it and they have talked with you enough to know you want quality work and that you are willing to pay for it. Chances are they will do an excellent job.

This is the way we do it in the contracting business; this is how I've done it for years. I drive around and look and shop job sites and when I find good workmanship and decide that the workmen are ones I can talk to, reason with and work with, I hire them. You should do the same. *Stroke* these people, let them know how much you appreciate their commitment to quality. I don't mean to B.S. them, I mean be sincere with your compliments.

GETTING BIDS

There are carpenters, for instance, who work hard all day and when they come home and eat, their workday still isn't over. They have to sit down, not in front of the TV but at a desk or table to write up bids. Their headache with dealing with contractors or those who are self-contractors is on par with the trouble you think you have with them. Some home owners are impossible to deal with. That's why I feel this book will help everyone, and I give you both sides.

If you're a homeowner about to build but one house, these carpenters, electricians, framers, cabinet makers, etc., have to contend with your not knowing what you want. You ask a bunch of stupid questions, your spouse has ideas, you want this there and that here and don't want to pay for it, you want something built for less than the cost of material and what you thought you wanted yesterday you no longer want today, and so forth.

This workmen gets no money for these bids and of course they get work. But it is, for the most part, an unrewarding part of the business. They have to talk with suppliers to get current prices, they have to make calls or visits, then get back to you. In the meantime, they have other jobs to finish. Oh I know, it's a part of every business, you say, and it is. Yet, it's no fun.

And, if a person has worked all day, they are tired. Going home to do more paperwork is not their priority so they might rush through a bid and then, nobody's happy.

To protect yourself against a shoddy bid, ask for a written bid, called a proposal. It has to be spelled out in detail as to what this person will do, how long it will take, and how much they will charge. They'll need a complete set of good drawings to do this.

You have to have it written, in detail, what they will supply or not supply, the total cost, their insurance and what type they have, their warranty and the payment schedule. On a plumber (or anyone) you have to have it written if they supply fixtures, water heaters or just connect it. Remember, "A short pencil is better than the longest memory."

For example, let the plumber list the charges on these various pieces of equipment that he supplies. You don't care about the pipes and connections but you do care about various heaters, commodes, and fixtures. If your plumber gets them you'll be charged for the appliances plus the time they took to pick them up.

You might be able to get them cheaper on your own at places like Home Depot, Builders Square or Wal*Mart and *see* them as they are. These are the same places they might shop and raise the price so why not do it yourself? There are, however, specialty items that might not be available at these retail outlets so ask the plumber about them and have them explain the difference in quality and price.

With an electrician, you might want to get your own lights or fans or bathroom fixtures, kitchen lights, etc. Yes, even not being your own contractor is still some work but far less than having to do it all!

If a proposal you get is hand-written, make certain you can read it. I prefer proposals that are typed. I've found that incomplete bids result in incomplete work and sloppy bids result in sloppy work. It's all part of doing business.

Don't go along with a verbal bid, or "We'll get to that later." No deal! Don't go for it! We call these people and type of bids "80-20-100" which means they'll expect 100% of the money and do only 80% of the job and leave the extra 20% for you or somebody else to do. Get it all written down!

THINGS TO LOOK FOR IN A PROPOSAL

1.   See how detailed it is; is it complete?

2.   Is it neat and self-explanatory? Is it not jumbled and comprised of terms you don't understand? And filled with half sentences that only confuse you?

3.   Have them list their warranty, terms and time limits.

4.   Breakdown on supplies and material—and labor. Will they cleanup or just finish, get paid and flee? Some do.

5.   Cost breakdown. Remember, your bank pays you so much as the job progress so blame it on the bank. Ask for a payment schedule and follow it as best you can. These guys can't wait until the entire job is done, they need money too. They will need

several draws. They need to be explicit with the work that is completed and when these draws are to be made and you need to make certain that the work they say is completed is, in fact, completed.

6.   A Certificate of Insurance is necessary to accompany the bid. Don't let the sub-contractor or worker supply you with this certificate of workmen's comp or liability, it must be mailed by the insurance company or agent of that company directly to you with your name and address on it! If a supplier attempts to give you a copy of this, it is no good whatsoever! Red flag! Check him out with his bank, supplier or the BBB.

A copy of a Certificate of Insurance can be old or a premium not paid, and the policy canceled! And you won't be covered. By having it mailed to you from the agent or insurance company, the moment the premiums aren't paid, it is their duty to notify you. Lot's of angles to look out for, aren't there?

| ACORD. CERTIFICATE OF INSURANCE | | | | | DATE (MM/DD/YY) 07-05-94 |

PRODUCER
MORGAN INSURANCE COMPANY
10777 NORTHWEST FREEWAY #350
HOUSTON, TEXAS 77092
(713) 682-0353

THIS CERTIFICATE IS ISSUED AS A MATTER OF INFORMATION ONLY AND CONFERS NO RIGHTS UPON THE CERTIFICATE HOLDER. THIS CERTIFICATE DOES NOT AMEND, EXTEND OR ALTER THE COVERAGE AFFORDED BY THE POLICIES BELOW.

COMPANIES AFFORDING COVERAGE

COMPANY A   R.L.I.

COMPANY B   T. W. C. INS. FUND

COMPANY C

COMPANY D

INSURED
IDEAL ROOFING
17411 VILLAGE GREEN DRIVE
HOUSTON, TEXAS 77040

THIS IS TO CERTIFY THAT THE POLICIES OF INSURANCE LISTED BELOW HAVE BEEN ISSUED TO THE INSURED NAMED ABOVE FOR THE POLICY PERIOD INDICATED, NOTWITHSTANDING ANY REQUIREMENT, TERM OR CONDITION OF ANY CONTRACT OR OTHER DOCUMENT WITH RESPECT TO WHICH THIS CERTIFICATE MAY BE ISSUED OR MAY PERTAIN, THE INSURANCE AFFORDED BY THE POLICIES DESCRIBED HEREIN IS SUBJECT TO ALL THE TERMS, EXCLUSIONS AND CONDITIONS OF SUCH POLICIES. LIMITS SHOWN MAY HAVE BEEN REDUCED BY PAID CLAIMS.

CO LTR	TYPE OF INSURANCE	POLICY NUMBER	POLICY EFFECTIVE DATE (MM/DD/YY)	POLICY EXPIRATION DATE (MM/DD/YY)	LIMITS	
A	**GENERAL LIABILITY** X COMMERCIAL GENERAL LIABILITY CLAIMS MADE X OCCUR OWNER'S & CONT PROT	CGL 00365196	04-21-94	04-21-95	GENERAL AGGREGATE	$ 1,000,000.
					PRODUCTS-COMP/OP AGG	$ 1,000,000.
					PERSONAL & ADV INJURY	$
					EACH OCCURRENCE	$ 1,000,000.
					FIRE DAMAGE (Any one fire)	$
					MED EXP (Any one person)	$
	AUTOMOBILE LIABILITY ANY AUTO ALL OWNED AUTOS SCHEDULED AUTOS HIRED AUTOS NON-OWNED AUTOS				COMBINED SINGLE LIMIT	$
					BODILY INJURY (Per person)	$
					BODILY INJURY (Per accident)	$
					PROPERTY DAMAGE	$
	GARAGE LIABILITY ANY AUTO				AUTO ONLY - EA ACCIDENT	$
					OTHER THAN AUTO ONLY: EACH ACCIDENT	$
					AGGREGATE	$
	EXCESS LIABILITY UMBRELLA FORM OTHER THAN UMBRELLA FORM				EACH OCCURRENCE	$
					AGGREGATE	$
B	**WORKERS COMPENSATION AND EMPLOYERS' LIABILITY** THE PROPRIETOR/ PARTNERS/EXECUTIVE OFFICERS ARE: INCL EXCL OTHER	Q19117	04-12-94	04-12-95	STATUTORY LIMITS	
					EACH ACCIDENT	$ 100,000.
					DISEASE - POLICY LIMIT	$ 500,000.
					DISEASE - EACH EMPLOYEE	$ 100,000.

DESCRIPTION OF OPERATIONS/LOCATIONS/VEHICLES/SPECIAL ITEMS

CERTIFICATE HOLDER
TOM TYNAN
6615 APPLE VALLEY
HOUSTON, TEXAS 77069

CANCELLATION
SHOULD ANY OF THE ABOVE DESCRIBED POLICIES BE CANCELLED BEFORE THE EXPIRATION DATE THEREOF, THE ISSUING COMPANY WILL ENDEAVOR TO MAIL _10_ DAYS WRITTEN NOTICE TO THE CERTIFICATE HOLDER NAMED TO THE LEFT, BUT FAILURE TO MAIL SUCH NOTICE SHALL IMPOSE NO OBLIGATION OR LIABILITY OF ANY KIND UPON THE COMPANY, ITS AGENTS OR REPRESENTATIVES.

AUTHORIZED REPRESENTATIVE
GIL MORGAN, PRESIDENT

ACORD 25-S (3/93)   © ACORD CORPORATION 1993

Figure (a) Certificate of Insurance.

Proposal — Page No. ___ of ___ Pages

Ideal Roofing, Inc.
17411 Village Green
Houston, Texas 77040
(713) 896-1122 • Fax (713) 937-7017

N° 20188

PROPOSAL SUBMITTED TO: JOHN DOE	PHONE Hm # 555-1212	DATE 7-15-94
STREET: 1234 ANY STREET	JOB NAME: T/OFF 1 & REROOF	
CITY, STATE AND ZIP CODE: YOUR TOWN, USA 09000	JOB LOCATION: SAME	

We hereby submit specifications and estimates for:

Tear of ___1___ roof(s); Recover ___0___. Remove all debris from property.
Install ___N/A___ decking over 1" x 4" lathes.
Install 28ga. drip edge. Apply 15# felt. Double layers in valleys.
Install ___3___ code lead jacks, ___2___ 1 1/2", ___0___ 2", ___1___ 3", ___0___ 4".
Install ___2___ starcap vents, ___1___ 4", & ___1___ 8".
Install ___1___ code base Flashings, ___1___ 3", ___0___ 4", ___0___ 5".
Install ___25___ lineal ft. 8" x 8" step flashing.
Install ___15___ lineal ft. cut in counter flashing.
Install ___10___ lineal ft. base flashing.
Install ___N/A___ metal pans, sizes ___N/A___
Re-roof using ___G.S. FIREHALT 25 YEAR___ shingles.
Use ___20 YEAR___ shingles for ridge and starters.
All shingles to be hand nailed ___4___ nails per shingle 1 1/4" nails.
Install 1-GAR Air Hawks 3 NEW Turbos ___N/A___ Power Turbos.
___TEN___ Year warranty on all workmanship.
Includes magnetic nail sweep of yard.

Roof price as stated above $ 2,995.00

OPTIONS (Additional charges apply to items listed below)

Up grade to "Ideal Roof Package" $125.00;
Kool ply decking ___N/A___; Remove 1" x 4" lathes ___N/A___; 30# felt paper $112.00;
3" x 1 1/2" 26 ga. DL galvanized drip edge $82.00; Prepainted $115.00;
___35___ Lineal ft. of ___NAME BRAND___ Ridge vent $132.00;
Soffit intakes $90.00; Plastic in attic ___N/A___; Build chimney cricket $125.00; Attic clean out ___N/A___;
Replace rotten decking at rate of $1.50 per sq. ft.
Replace rotten fascia at a rate of $4.00 per lineal ft.

NO DEPOSIT REQUIRED. PAYMENT DUE AT COMPLETION.

Pay upon completion unless agreed otherwise in advance.

All accounts due and payable in Houston, Harris County, Texas. A service charge at the highest legal rate may be charged on past due accounts and reasonable attorneys fees if account is placed for collection

Price of Options $_____
Total Price $_____

Authorized Signature ___Rodger Brogdon___

Page Number 01 of 01 Pages

Note: This proposal may be withdrawn by us if not accepted within __30__ days

Acceptance of Proposal - The above prices, specifications and conditions are satisfactory and are hereby accepted. You are authorized to do the work as specified. Payment will be made as outlined above.

Signature _____

Date of Acceptance _____   Signature _____

Figure (b) Type of Proposal.

Proposal

Page No.   of   Pages

This firm endorses
The Better Business Bureau
Code of Consumer Rights

Ideal Roofing, Inc.
17411 Village Green
Houston, Texas 77040
(713) 896-1122 • Fax (713) 937-7017

No. 10555

PROPOSAL SUBMITTED TO
JOHN DOE

PHONE   555-1212

DATE   7-15-94

STREET
1234 ANY STREET

JOB NAME
FLAT ROOF REPLACEMENT

CITY, STATE AND ZIP CODE
YOUR TOWN, USA 09000

JOB LOCATION
SAME

DATE OF PLANS

JOB PHONE

We hereby submit specifications and estimates for:

TEAR OFF EXISTING FLAT ROOF. REPLACE ROTTEN DECKING AT A
RATE OF $1.50 PER SQUARE FOOT (EXTRA CHARGE). INSTALL CANT STRIP
AT ROOF WALL TRANSITION ON UPPER ROOF. APPLY 23# FIBERGLASS BASE
SHEET USING SIMPLEX FASTENERS. ON UPPER ROOF INSTALL 2 NEW 4" DRAINS
AND 2 NEW 4" x 3" "M-WELD" MODIFIED BITUMEN SCUPPERS. INSTALL
4 - 8" (0/12 PITCH) VENTS AND 1 - 14" BASE FLASHING. MOP DOWN 2
PLIES OF G.S. PLY 6 PREMIUM PLY SHEET USING 25-30# PER SQUARE TYPE III
HOT ASPHALT INTERPLY MOPPINGS. ON LOWER ROOF INSTALL SINGLE PLY EDGE
METAL. MOP DOWN 1 PLY G.S. GMS MODIFIED BITUMEN FLAT ROOF SYSTEM
USING 25-30# TYPE III HOT ASPHALT. ALL METAL TO BE PRIMED WITH
#207 PRIMER.

REMOVE ALL DEBRIS CREATED BY JOB.
ALL WORK TO BE DONE ACCORDING TO G.S. SPECIFICATIONS.
PROVIDE A FIVE YEAR WARRANTY ON ALL WORKMANSHIP.

We Propose hereby to furnish material and labor - complete in accordance with above specifications, for the sum of:

THREE THOUSAND FOUR HUNDRED SIXTY & 00/XX DOLLARS --------($3,460.00)

NO DEPOSIT REQUIRED. PAYMENT DUE AT COMPLETION.

(PAGE NUMBER 1 OF 1)

All accounts due and payable in Houston, Harris County, Texas. A service charge at the highest legal rate may be charged on past due accounts and reasonable attorneys fees if account is placed for collection

Authorized Signature   Rodger Brogdon

Note: This proposal may be withdrawn by us if not accepted within 30 days

Acceptance of Proposal - The above prices, specifications and conditions are satisfactory and are hereby accepted. You are authorized to do the work as specified. Payment will be made as outlined above.

Signature _____

Date of Acceptance _____

Signature _____

Figure (c) Another type of Proposal.

COMPARING BIDS

There are things to look for when taking bids. If you get three bids from plumbers and one is far cheaper than the other, naturally you'll look at the less expensive one first. Red flag! Check the details!

If it's detailed, it will show the type of water heater, for instance, check to see if it's the same brand and capacity as the others. Look to see if the fixtures are comparable. Does he carry insurance and does he warranty his work? I've heard about some of these "low bid" people who just vanish before the job is completed and move on to another area or town and give more low bids. You really have to know some things whether you build yourself or hire a contractor.

Sometimes these differences are difficult to see, unless you read the bids carefully. A water heater, for instance, might be the same brand but a different capacity, such as a 30, 40 or 50 gallon. Or, it might be the same brand and capacity but a low-grade model. These are things you have to know and (again) you can't know it all. The best way is to ask! Ask the person giving the bid, ask the one who gave you another or call and ask me!

There are so many different kind of bids to consider. The ones that are out in left field, you know, either ridiculously low or absurdly high, just throw away. When you find three bids that are reasonably close to each other, pick through those and decide.

This is when appearance and personality become so

important. Choose the one who sounds the best, the one who talks plain, looks you in the eye and answers your questions without throwing in scare tactics or detailed information you could care less about. *Anal retentive* people, I've found, (bless their intelligent souls) will share all sorts of information with you that you neither care about nor understand. Ask them what time it is and they'll tell you how the clock was built!

Smart contractors know how to talk to a person without insulting them or boring them to tears with the tensile strength of this, the coefficient of linear expansion of that and British Thermal Units of something else.

CHOOSING SUBS

If the guy is neat, has recently shaved, smells reasonably well and answers your questions satisfactorily, go with him. In my first book, **Home Improvement**, the first chapter deals with selecting a repairman and the same holds true with selecting a contractor or sub-contractor.

One method I've found effective is to find out **where they buy their material.** Call the supply house and ask for the billing department; find out if they pay their bills on time. If they have a good track record, chances are they run a reputable business. If they pay late or owe the supply house money, chances are high that he will do the same to you. Pass them up and go to the next bidder.

If they refuse to give you the name and telephone number of their suppliers, **Red Flag!** Pass them up and save yourself time. No need to call the suppliers, their refusal to give you a list of where they buy their materials is all the answer you need.

A second way to check is to ask for a list of builders in the community they've worked for in the past. Then, call these builders! Ask them questions like:

(1)   Were they happy with their quality of work?

(2)   Would they hire them again?

(3)   Do they show up on time?

(4)   Any drug or drinking problems?

(5)   Is their crew made up of the same people all the time?

(6)   Do they try to hit you up with cost overruns or unnecessary changes?

When you get the builder or contractor on the phone, ask specific questions and don't waste a lot of the builders or contractors time; they have work to do. Bring this book with you and ask the questions I specified.

Again, I am not saying that the building industry is comprised of all crooks and charlatans, hardly. There are, however, those who sneak in and "clip" people for whatever they can and move on. That's why books like this, programs like mine and courses that are available are important.

Nobody expects you to know it all because we all have different lines of work. Mine just happens to be building and dealing with builders, suppliers and sub-contractors and in knowing what to look for (and look out for) on bids and work promised. Get it down in writing and if you aren't certain what this or that means, ask! Then, get that in writing!

Qualified, reputable sub-contractors will appreciate your thoroughness. It will prevent all the problems from popping up later on. An educated consumer is the contractors best and most revered customer.

A GOLDEN RULE

In defense of the contractors, many people *shop* bids; they get a bid and then give a copy of that bid to someone else to bid lower then go to someone else with that lower bid and word gets around.

Sub-contractors work for any number of different contractors. When they find out what you are attempting, they're human and might go after you in some fashion like giving you a ridiculously low, low bid and building your hopes up, then canceling before they sign anything.

Some might substitute inferior materials or shoddy workmanship but most will just pass you up and tell others about it. That's when you'll really have trouble getting a legitimate bid. I've seen that happen a time or two, in fact, I've done it a time or two myself. Indulge me a moment for this short story. I like telling it.

There was this guy in Florida who shopped his bids to several contractors. I was about the third to bid, and by the time he had gone to the fifth contractor, we all knew about him. So I went back and bid maybe 20% under the lowest bid. I had all the business I could handle at the time anyway and I had a few minutes for fun and games to teach this guy a lesson.

I didn't like him when he first came in anyway and probably wouldn't have done his job if he had paid me 20% more than the top bid. There are too many nice people in the world and life isn't worth the hassle.

Well, I spent maybe 10 hours with him but it was cheaper than going to a rock concert, ball game or playing pool . . . and, more fun. I saw his eyes open wide when we talked about what he wanted, a look like a prospector who just hit the Mother Lode. His sheepish grin of conquest, feeling he had a "live one" on the hook made me chuckle inwardly because I knew that I had him!

The thing was, that he wasn't a *neophyte* in the building business. I found out through our contractors grapevine that he built two or three homes each year to sell and that he had beaten several hungry sub-contractors to where it cost them to do the job, a few he didn't pay at all. He told them to sue, knowing full well that they trusted his supposed naivete and didn't have the funds to fight him. This guy was a menace!

He wasn't content with good work at a good price, he wanted to *beat* somebody. It was the way he operated. I hope he doesn't have any kids; I'd hate more like him running around in the future to prey on my sweet, honest children.

Anyway, I didn't gyp him out of any money nor did I install faulty material, I was having a good time and wouldn't come down to his level. But I did get him excited and I did take up a lot of his time telling him of the special deals I had where I could build quality for less. He liked that.

The day I called and told him that my deal had fallen through, I wish I'd been there to witness his expression. There was a dead silence on the phone for maybe a full minute and I swear I heard a sob, at least a deep breath, maybe a gasp. And the fact was that nobody in the entire city of Ft. Lauderdale would work with him.

I don't know where he is now or what happened to the house other than the lot was up for sale and he didn't build. From time to time, many of us good guys show the get even side when dealing with a jerk.

> My Golden Rule is never shop bids with various subs using the lowest bid as bargaining power. You get what you pay for.

One version of The Golden Rule is "The one who's got the gold, rules!" I hope you laughed at it; I did.

I'm not writing a book for contractors or sub-contractors

to make them aware of what some of the things they have to deal with because they already know. There are some difficult people out there who don't know what they want and sometimes, those that do, want to pay so little that the contractor or sub can't make a profit.

And there are those who want to change things once they're in and that's natural but not a lot of things, okay self-builders or contractors? Don't try to mooch these changes, a change costs money. An upgrade costs money, a little thing here or there costs money and somebody has to pay. If you want it, be prepared to pay for it.

Uh Oh! Here I go now talking like a contractor. I'm not a contractor any longer, *per se*. I still build a house once in a while to break the monotony and to see if I'm as good as I think I am. Actually, I love building and we all enjoy doing what we think we're good at, don't we?

I like using all that I've learned to build a really fine house. You will too. The feeling I get when I finish a house is (I imagine) like a mountain climber reaching the summit or maybe the thrill of winning an Olympic gold medal. You'll know what I mean when you finish your house. Call me and tell me if I'm wrong.

Just play by the rules and everything will come out okay. Learn what to do by reading, studying, attending classes on building then write it down, ask questions and begin. The beginning is a bit frightening but the end is "Oh so sweet!"

When I build, I just get a high that is inexplicable. When I get up and put on those work clothes, I just feel alive and ready to handle any problems that come my way. It makes me feel vital! As a builder you are creating something special for yourself and your entire family.

Make it all fun. When your home is completed, you'll feel as if you just won an Olympic gold medal and the band is playing the Star Spangled Banner as you stand on the highest podium. Then, you'll be able to understand how I think and feel. There's nothing like it.

Chapter 9

SEQUENCE OF EVENTS

If, after all of this, you still plan to build your own home, I'd like to take most of this chapter to tell you the **sequence of events** you must go through before you begin building.

We've broken down all the different things that must be done before you break ground but we never talked about the order in which they need to be handled. I'm doing that now.

THE PRE-PLANNING STAGE

You literally have to sit down and determine the goal you plan to reach. There are some questions you have to ask yourself. Do you own a house? Have you owned a house? Do you want a bigger house? Do you want the same house remodeled to where it's almost new? Do you plan to remodel this one to were it is neat, sell it and build someplace else? Will you fix this house up to rent? Will your new house be in the city or a cabin in the country?

These questions are important to ask yourself; the answer is even more important. The answer will dictate your next move. My father, for instance, collects and builds old Model A Fords. His house could be a bedroom, bathroom, a kitchen with a hot plate and a 6-car garage! Once you have the answer, you are on your way.

NOW YOU MUST DECIDE:

(1)   Whether you want to be your own
         contractor?
(2)   Can you be your own contractor?
(3)   Do you hire someone else?

You alone have to make this decision because it means everything to the rest of the plan. Let's assume you have decide to build a new home and sell your old one, after you do some touch up work on it.

The average time it takes to build a home, I mean from the very day you decide to buy land or build on land you already own, is two full years!

People are under the impression that it takes three months to build a home and this is wrong! Maybe a tract builder can build in three months because you have little or no say in the matter and everything they do is done except the actual building of the house.

PLAN OF ACTION

(1)   The first person you call is your architect/de-signer to get the plans and specifications moving along. This can take anywhere from 4 to 6 months!

Make certain you get at least 8 sets of plans from your architect and as you hand these plans out, request that you get them back from those who bid the jobs.

One set will go to the air conditioning people, plumber, electrician, framer, and foundation and when they give them back to you, that's 5 sets. As you get bids from these various people, get the plans back because when you decide on who gets the job, they'll need these plans to begin the bid.

Be careful not to give up your set of plans that you use for your permit to anyone giving you a bid. There will be times when a set of plans or two are not returned and if you give up these precious permit plans and someone from the inspectors office wants to see them, you're in trouble and there will be a delay.

(2)   If you aren't going to act as your own contractor, you must hire one. Now you have to choose one. How you choose one is described in the previous chapter.

(3)   Either your contractor has to put together a total bid of the cost of the job or if you're acting as your own contractor, you must do it.

This is when you solicit bids if you're acting as your own contractor but if you hired a contractor, they solicit these bids and give you the total cost, including their fee.

Now, you're almost ready and you haven't made a mistake because you made the decision what to do, you have your plans, you've hired a contractor, and you have your total cost. Now, it's time to approach your banker.

(4)   Prepare your loan package and take it to your banker. I'd suggest you shop your loan package to at least

three banks, then decide on the bank (and banker) you feel you could best work with.

> If you're bidding the project yourself, total up the highest bids and submit those with your loan package. I'm not saying to cheat your bank but maybe fudge a bit; they expect it. If you don't, it's a mistake because you always need a slight "cushion". Remember, you don't necessarily have to use the money but it's there if you do need it.

I think it's sort of a game with builders and banks. The banker knows you have added costs and that lumber prices vary. Most assume you are using the highest priced bids on materials and labor and all you want is a full 80%-of-cost loan.

When the smoke clears from the actual house being finished, you have to think about cleanup crews, maybe a fence, perhaps landscaping for the front yard that includes an extra tree or two, full sodding instead of sprigs, hedges, bushes, maybe an underground sprinkling system. And what about the backyard landscaping?

Most first-time builders forget these extras, but they add up. The bank doesn't care; they have more money than you, they collect interest on what you borrow, and you're going to pay them back anyway. Too, it's easier for you to pay over a 25 or 30 year period than it is to come up with an extra five or ten thousand out of your pocket that you didn't bargain for.

FINAL CHECKLIST

(1)   Pull your permits and make certain you're in compliance with all the rules and regulations in your area.

(2)   If you're acting as your own contractor, begin scheduling your sub-contractors and material suppliers. If you hired a contractor, they take care of this.

In scheduling sub-contractors, do this at least 7 days in advance so your job won't be held up. In fact, contact them as soon as you make up your mind and begin making a schedule.

There is a lot of difference in building a new home as compared to the relatively smaller job of remodeling a bathroom. Now, let's get his project off the ground!

Remember, no matter how small the project, in order for it to run smoothly, you still need to go through your planning stages. A bathroom remodeling is a great job to sink you teeth into for a first-time contractor. It has the same problems as an entire new home, just smaller.

It has plumbing, electrical fixtures, floors, cabinets, hardware, wallpaper, paint and tile all in one small area. If you make a mistake here, better than on an entire house.

REMODELING A BATHROOM

When a homeowner remodels a bathroom it's usually because the tile or fixtures are outdated and they want new *stuff.* Let's take a typical bathroom to be remodeled and walk through it one step at a time. I don't mean to insult anyone with some of these basics but I wasn't as smart as some of you when I did my first building job and I'd have appreciated an a—b—c plan.

THE SEQUENCE IN REMODELING A BATHROOM

1. Draw or have plans drawn.
2. Do a total cost breakdown to make sure you can afford it.
3. Gut it. Tear out walls and floors.
4. Haul off the mess.
5. Buy new tub, commode, lavatory, light fixtures.
6. Framing for walls and ceiling.
7. Insulation for outside walls.
8. Plumbing.
9. Electrical.
10. Replace sheetrock/backer board.
11. Tape and float sheetrock.
12. Contact cabinet makers.
13. Consult painters.
14. Do tile work.
15. Install mirrors, shower doors and hardware.
16. Hang wallpaper last.

> If you do it yourself, the material and supplies are paramount. In your plans, on your takeoff, get everything you need to do the job, AT the job! It is a pain in the neck and a waste of time and money to have to go back and get this, go back and get that.

The remodeling of a bathroom can be fun. The man usually does the grunt work, the woman enjoys choosing the paint and wallpaper. If you get help from a friend or relative who has done this type of stuff before or can hire that handyman, it makes it even easier.

Common sense often comes into play. You can make a game of this remodeling business by getting the entire family involved. Each of you get a sheet of paper and write down, alone, what you think has to be done. Then bring the papers out and compare notes. Then grab this book and check off what you might have forgotten.

You can do a kitchen or garage the same way. You can add a room, even build a house if you follow this sequence and (again) if you have the time! Time is the most important factor on deciding whether to do it yourself. Well, maybe having the money to do what you'd like is first but, *time* is a close second. If you have the money to do it and no time, have someone else do it. If you have the money . . .

Reminds me of a song. Remember the one that goes, "If you've got the money, honey, I've got the time?" I don't think they were talking about remodeling a bath or kitchen but, same principle except you have to have the money and the time to remodel anything and do it yourself.

If you do remodel on your own and this book doesn't tell you what you need to know, call my program, especially if you've already bought my book!

This book is to get you started on the right path to building and to help you gather the courage to try. It doesn't tell you how to build; you learn that from classes, reading, watching and doing. I can only show a horse where the water is and encourage him to go the trough, I can't drill the well.

Chapter 10

MOST COMMON MISTAKES

These are the errors many people make when attempting to build their own homes. We all want to save money but by trying to avoid the necessary steps will only cost more money.

NOT PULLING PERMITS

Many think they can "beat the system" and not obtain any building permits. "Nobody will ever know. I don't need those city people involved. I don't want them to know what this cost me because they'll raise my taxes."

This is an error. Get the permits because the consequences can be painful. If they find out and send someone out to inspect, they could fine you on a daily basis until all things are taken care of. You might think, "If they find out, I'll just run down and get a permit," but it isn't that easy. Once you've been found out, they'll dog you like crazy on whatever you try. You'll be fined, delayed and legally harassed to the best of their ability. Do it right and get the permits!

INSUFFICIENT PLANS AND SPECIFICATIONS

"I don't need to spend all that money on plans. I can just take on the problems as I go along."

Trust me, ladies and gentlemen, get the plans. For a house that might cost $75,000 you might expect to pay $4,000 or $5,000 for a set of good plans to maybe $10,000 on a $150,000 to $200,000 home.

My experience has been with many, many homes and a perfect example is a simple swing set for kids. Just try putting one of those things together without looking at the directions. Then take that and multiply it by about a thousand and you'll get the general idea.

Too, the plans and *specs* are your legal documents for your contractor. These plans are done by architects, designers and/or engineers. Without these plans, your contractor can do as they please. These plans tell the width and length of wood, the grade of wood, type of fixtures, they tell everything. If your contractor doesn't follow these plans, they are your proof in a lawsuit.

NOT DOING A DETAILED COST ESTIMATE

If you have no idea of the cost of your material and labor, it's the same as taking a limited amount of money with you for a vacation without knowing the price of the airfare or hotel. You might say, "I've got $10,000 and I can finish it with that."

A major error! You could end up with one wall missing from your bedroom or not enough money to put a door on it to keep the kids locked out. Or, you'll have to borrow money from your brother-in-law who will remind you (and everyone else within earshot) of it daily!

You need plans and must have a detailed cost estimate before you go to your bank for money. If you don't and the bank loans you money on your signature or on a second lien on your house and you run out of money, then where do you go? To your brother-in-law?

GOING WITH THE LOW BIDS

"Oh, this guy is great. He's 30% less than the other two bids."

Many people who bid low aren't necessarily crooks or want to gyp you, they might just be unskilled and stupid and want to build for their quote but find that they cannot. So what happens? They leave an unfinished job and you are stuck with it.

Get three or four bids and go with the contractor you like best who is in the ballpark of the other bids. If you find one that is drastically lower, a red flag needs to be hoisted and you can start asking questions or that person just might be giving a dozen or so bids in that area and complete half the jobs and move on.

When I say these things, I am by no means saying that contractors are cheats or sleezeballs, the majority of contractors are reputable, knowledgeable people. But it's the one here and the one there that we hear about most. These stories tend to stick. Just be smart and not in too much of a hurry to choose. Get three or four bids - then make your decision.

USING SUBSTANDARD MATERIALS

Another big mistake is by trying to use less expensive, inadequate material for a job. If you have old wood that is questionable, put it to rest. Some friend might offer you some wood, "Say, I've got this old cedar, it's a little wormy but if you spray it, it will hold a nail. It will save you money." Remember, "A penny wise and a dollar foolish." It is far better to do it right the first time than to have to tear it out and do it again.

PROJECT TOO BIG FOR YOUR SKILLS

We talked about this in the chapter of being your own contractor or hiring one. If you don't have the skills to do the job, don't even try it. Don't let your ego force you into a dumb mistake.

After years of building, remodeling and watching these various carpenters work, I have as much admiration for them as I do for brain surgeons. These guys are fast and they are good. If you don't think so, take some time off to look at some carpenters work. Or watch a guy laying brick or setting floor tile. Look at any of these technicians who are expert at what they do. It takes years and years and lot's of practice. If the job is too big, don't try it, call a contractor.

The first project of size that I'd recommend you try is maybe remodeling a bathroom, or maybe adding a room. Even if you want to add a room, make certain you know everything about it, all the pitfalls. You might not fail or rather, you might not fail big, but it could end up costing a lot more or you might do shoddy work which will detract from the looks and certainly from the value of your home.

MAKING TOO MANY CHANGES

Woe is me! Woe is me! This happens to those who are doing the work themselves and more of a headache is making changes with a contractor, especially if I'm the contractor.

I've had customers, I swear, whom I feel stayed up all night thinking of things to change that would aggravate me and really mess up truly wonderful plans that the architect drew. Still, it is your prerogative and if you like it better one way, do it their way ! After all, it's their house and they are the ones paying for it.

I'd prefer, of course that they do their utmost to get it the way they want it in the plans. Still, these things happen, and I'm not knocking it; it's happened to me in building my own house.

"Let's go with a better grade of marble and not the original tile." Or, "I think I'd like a bigger bathtub or take off that trim and put this new kind on." Or, "I'd like a bidet right there (they point to the plans) right next to the commode".

The problem with the last suggestion is that it won't fit, unless their left foot is in the bathtub when using the bidet or their right foot would have to be in the commode.

Some changes are expected but a lot of changes are a problem for a builder but a larger problem for your pocketbook. Try to decide **before** you start building. It will save us both, especially you. I can tear out anything and rebuild it but remember, it will cost! It all adds up friends. An extra $200 here and $50 there and $400 here, when tabulated, can be substantial.

Yet another problem with these changes is that they aren't in the package from the bank of specifications and you'll have to pay cash for those. They weren't covered in your loan. And, you'll need this cash when moving into a home because there are a lot of other things you'll need. For instance, curtains, drapes, fabrics, furniture, light bulbs, telephone and television hookups, upgrade of carpet etc., etc.

UNDERESTIMATING CASH RESOURCES

Other than the bank loan, you'll need cash for the changes you will inevitably make. Banks, for instance, only lend you 80% of the total cost of the project, and give you cash draws on your work as it progresses.

They'll give you a specified amount of money on your framing and then on your roofing. You might need cash for when an emergency occurs. For a $100,000 home, I say set aside maybe 10 or 15 thousand ready cash for these emergencies. Chances are you'll be able to recoup this on future bank draws down the road a bit, but you need it now!

LISTENING TO BAD ADVICE

I can't begin tell you how many jobs I've been on when a customer came up and said, "Listen, my dad is from Alaska and he said when he used to do this he'd put this on first or not do that or he said you are doing things not necessary to do and that ups the price and, of course, your profit."

Maybe 40 years ago or up in snow country it was the way to do it but things change. I'd listen to the advice and suggestions and if it sounds reasonable, consider it. If not, smile, thank them and go about your work.

You see, one of my goals is to clear up the mysteries of building a home. Through my books and over my radio and television programs, I try to make this understandable for everyone. This, of course, helps the contractor but it also helps you save money, time and aggravation.

I realize it sounds self-serving but, use me! I studied building, I did it, I do it, and I still study and research. I built homes and remodeled homes that cost from $40,000 to well over a million dollars. Call me during my radio program, call me or watch me over television and buy my books. The books are cheap, the phone calls might cause you to wait a while (hopefully) but it costs nothing to call. I truly want to help you. And I will, save you money!

This just about wraps up this book. I know I mentioned time over and over again but I did it because it is "the" single most important element in successful building or remodeling. Good by and God bless.

Tom Tynan

TOM'S SECRET REFINISHING FORMULA

Materials: Mineral spirits (paint thinner)
                   Boiled linseed oil
                   #0000 Steel Wool
                   Clean rags (preferably soft cotton)

I learned this when I was 15 years old while working in an antique shop in Florida. They were masters at restoration and would not attempt to strip and refinish an antique, it loses its value. Try this on a small area first.

The products are relatively inexpensive and one treatment lasts a long time. It is an easy, gentle and excellent method in bringing life back into old wood such an paneling, cabinets or furniture. The mineral spirits cleans the grease and dirt from the surface while the linseed oil moistens the wood and blends any scratches in with the existing finish.

Mix one part linseed oil with two parts mineral spirits. Saturate a hand-size piece of this #0000 steel wool in this mixture and rub lightly, making certain to follow the grain pattern. Then, take the clean rags and wipe off the excess.

If the area you've just treated feels sticky, put a dab or two of mineral spirits on a fresh clean cloth and wipe the area down to remove the excess linseed oil. If the mixture starts to thicken, add some mineral spirits.

Other Books by Swan Publishing

HOME IMPROVEMENT . . . This is Volume 1 and the first book Tom Tynan wrote on home improvement. It has literally **hundreds of** homeowner's most often asked questions, **answered!** In but 3 years it is in it's **eleventh** printing . . . . $ 9.95

BUYING & SELLING A HOME . . . Volume 3 by Tom Tynan, and it gives you the "secrets" of buying as well as selling a home. It is so thorough, informative and innovative, that thousands of Realtors have bought it to give to their clients. By doing this, they attest to the fact that the book is extraordinary . . . . $ 9.95

STEP BY STEP *15 Energy Saving Projects* . . . Volume 4, is Tom's newest book. It is his best, he feels. Energy companies all over the United States are buying thousands of copies to give to their customers . This also, is a compliment to Tom's writing and knowledge . . . . . . $ 9.95

HOW NOT TO BE LONELY . . . If you're about to marry, recently divorced or widowed, want to forgive, forget or both, this is an excellent book to read. Candid, positive, entertaining and informative, written by best-selling author, Pete Billac. It is a fun book to read with answers that will help you get a date or a mate. It tells you where to find them, what to say and how to keep them. (Over 3 million copies sold) . . . . . $ 9.95

HOW NOT TO BE LONELY *TONIGHT* . . . The sequel to the book above, aimed at the *MALE* reader. Other than being courageous and strong, smart women want their man to be sensitive, caring, and understanding. "The" book to give to your man. Or, for men who really want to learn what turns the modern woman on . . . . $ 9.95

NEW FATHER'S BABY GUIDE . . . Another best selling book by Pete Billac. The **perfect gift** for ALL new fathers. There is not a book for new fathers quite like this one! Tells (dummy dad) about Lamaze classes, burping, feeding and changing the baby plus 40 side-splitting drawings by athlete/cartoonist Cash Lambin. Most of all, it tells dad how to **SPOIL** mom! GET IT for that new daddy! . . . . $ 9.95

YOUR FRONT YARD . . . A fun book of information by KTRH Gardenline co-host John Burrow, tells about plants, trees, grass, pesticides, fertilizer, everything you need to know to win have "the" best looking yard in your area. John makes this book for everyone to read. KTRH Garden of the Month contest . . . . $ 9.95

VEGETABLE GARDENING *Spring and Fall* . . . by John Burrow, is his newest book and tells everything a beginning gardener needs to know to grow a garden in the country, city, or in an apartment. A really great book! . . . . $ 9.95

ELVIS IS ON THE LOT . . . By Jim (Mattress Mac) McIngvale with Dave White, is the true story of how a couple invested $5000 and became millionaires by hard work, advertising, a plan, perseverance, and a goal. It has the formula about how **you can do it too!** It is a well written, excellent book to read and enjoy . . . . $ 14.95

HOW TO BUY A NEW CAR & SAVE THOU$ANDS . . . Inside information on dealerships and salespersons. This book by Cliff Evans, a former car salesman and general manager, really will save you thousands on your next new car purchase . . . . $ 9.95

REVERSING IMPOTENCE *FOREVER* . . . A truly great book written by two world famous urologists, Dr. David F. Mobley and Dr. Steven K. Wilson. This books tells MEN how they can REVERSE this problem but, woefully, men just aren't buying the book. It seem that men feel it's a terrible thing to mention whereas the terrible thing is **not doing anything about it!** The book has many drawings that explains how impotence can be reversed . . . . $ 9.95

We thought you might like this recent photo of Tom Tynan. It is on the cover of his latest book, STEP BY STEP.

TOM TYNAN is available for personal appearances, luncheons, banquets, home shows, seminars, etc. He is entertaining and informative. Call (713) 388-2547 for cost and availability.

For a copies of Tom's books - *HOME IMPROVEMENT; BUILDING & REMODELING; BUYING & SELLING A HOME: or STEP BY STEP (15 Energy Saving Projects),* send a personal check or money order in the amount of $12.85 per copy to: Swan Publishing, 126 Live Oak, Alvin, TX, 77511. Please allow 7-10 days for delivery.

**To order by major credit card 24 hours a day call:
(713) 268-6776 or long distance 1-800-TOM-TYNAN.
Delivery in 2 to 7 days**

Individuals who wish to contact Tom Tynan in regards to a building or remodeling question are asked to contact him through KTRH radio during his show. The number is 526-4-740.

★ ★ ★ ★ ★

Libraries—Bookstores—Quantity Orders:

**Swan Publishing
126 Live Oak
Alvin, TX 77511**

**Call (713) 388-2547
FAX (713) 585-3738**